The Value of Advanced Manufacturing Technology

The Value of Advanced Manufacturing Technology

How to assess the worth of computers in industry

J. S. Busby

Butterworth-Heinemann Ltd
Linacre House, Jordan Hill, Oxford OX2 8DP

 PART OF REED INTERNATIONAL BOOKS

OXFORD LONDON BOSTON
MUNICH NEW DELHI SINGAPORE SYDNEY
TOKYO TORONTO WELLINGTON

First published 1992

British Library Cataloguing in Publication Data
Busby, J. S.
Value of Advanced Manufacturing
Technology: How to Assess the Worth of
Computers in Industry
I. Title
670.285

ISBN 0 7506 0476 X

Library of Congress Cataloguing in Publication Data
Busby, J. S.
The value of advanced manufacturing technology: how to assess the worth of computers in industry/J.S. Busby.
p. cm.
Includes bibliographical references and index.
ISBN 0 7506 0476 X
1. Manufactures – Technological innovations. 2. Manufacturing processes – Automation – Economic aspects. 3. Computer integrated manufacturing systems – Economic aspects. I. Title.
HD9720.5.B836 1992
338.4'54–dc20

91–46318
CIP

Typeset by STM Typesetters Ltd, Amesbury, Wilts.
Printed and bound in Great Britain by Billings & Sons Ltd, Worcester.

Contents

Preface

This book is about the contribution that advanced manufacturing and engineering technologies make to the worth of industrial firms. It is concerned in particular with the appraisal of computers: with the process of understanding the nature of computer-based technologies, making predictions about their costs and benefits, and recommending sensible courses of action. This is partly a case of testing new systems against general economic yardsticks. And it is partly a case of testing them against the specific strategies that most firms follow in order to give a sense of direction to the changes they make in the way they conduct their business.

An appraisal is not just an opportunity to rule proposals acceptable or unacceptable. It is a process in which hazy ideas about promising technologies have to be clarified, and one in which a firm's managers can be explicit about the indicators of performance on which they want technologists to concentrate. It ought to be the process in which a technology's special properties are understood – properties that might make it difficult to introduce or operate. It is a good point at which to establish coherent lines of long-term development, and to begin spreading knowledge of a technology throughout an organization.

There are a number of difficulties associated with making decisions about technologies as new, complex, and wide-ranging as computer-based systems. Most of these are rooted in the problems of appraisal. Benefits are hard to quantify; costs are hard to pin down; uncertainty is hard to measure; experience and knowledge of the technology is expensive to acquire. To tackle these problems is to improve the way decisions are taken. It is to make better decisions about which new developments to start, and about which directions existing work should follow.

Economic appraisals are meant to inform, not automate, the decision–making process. There is never a guarantee that the information an appraisal contains will be enough to make good choices. A decision maker cannot be sure that it captures the full range of motivations, experience and insight that are to be found in any firm. Judgement therefore remains central to investment. But the fact that computer-based technologies are so different from their predecessors means that earlier ways of thinking about the worth of investments are now suspect. Understanding how to make decisions about advanced systems is therefore a matter of finding out how to make fully-informed judgements: judgements informed about the nature and the purpose of developments in information technology.

The first part of the book is concerned with laying out the context within which individual investments are assessed – with understanding the changing nature of manufacturing and the common characteristics of advanced technologies. It is only by looking at these long-term patterns of progress that it becomes possible to grasp fully the contribution a new system is likely to make.

The core of the book describes the yardsticks that can be used to express the value of a technology, and it discusses how the special characteristics of computer-based systems affect the way these are applied. These yardsticks are meant to capture both the immediate effects of undertaking a new project and its broader (and perhaps more tenuous) impact on other developments.

The final part is a discussion of practical issues – the influences on the appraisal process that are usually found in commercial organizations. They include the manner in which firms search for opportunities to apply advanced systems, how and why they might form strategies for doing so, and the methods they can use to assess technological risks. There is also an outline of the way an appraisal of new technology fits into the broader administrative activity – how it is connected with historical reporting systems and the use of budgets, for instance.

The book is for people who need to gain a good understanding of the way analytical tools can be applied to decisions about advanced manufacturing technology. The discussion is aimed at people who want a consistent and enlightened way of thinking about the economic issues that accompany both the introduction of new systems and their subsequent operation. This is meant to include managers and technologists in industrial firms, consultants and researchers.

The intention has been to provide a practical guide – but one that describes the underlying ideas in enough detail to make the down-to-earth calculations and tests understandable, consistent and convincing. Some of the tools are quantitative, but most can be used more informally. Some of them are relatively unproven, while others are well-established. But all of them rest on some basic reasoning, and to know how far the application of these tools can be taken it helps to trace this reasoning. The purpose of this book then is as much to explain this underlying structure as it is to give examples of the tools in use.

1 Introduction

He must behave like those archers who, if they are skilful, when the target seems too distant, know the capabilities of their bow and aim a good deal higher than their objective, not in order to shoot so high but so that by aiming high they can reach the target.

Niccolo Machiavelli *The Prince*

1.1 Opening remarks

Purpose

My intention here is to discuss the way in which the worth of investments in computer-based manufacturing systems can be assessed. This will partly be a case of examining a number of established models commonly applied to this type of work, and seeing how their properties are affected by the special characteristics of computer technologies. It will also be a case of looking at newer ideas that have not, so far, gained a widespread acceptance. The concern is both with fundamental questions (whether the models are consistent and rational, perhaps) and with practical questions – whether, for instance, they are easy to understand and cheap to use.

In a good many companies the appraisal process is a refined and stable one. Nonetheless, there are two compelling reasons for re-examining the approaches taken to appraisal whenever they are applied to new technologies. The first reason is that some of the rules that people use to make investment decisions do not represent very good generalizations of individual experiences. The observation that a firm's financial performance sometimes increases with the rate at which its machines are used has been embodied in rules that call for maximum utilization. Such rules are often applied when they are completely at odds with the intention of improving a firm's earnings – when machines are fully utilized only to produce defective products, for instance. The second reason is that the experiences on which current decision-making processes have been based are no longer representative of the problems and opportunities with which most firms now need to deal. At a time when technologies and organizations were broadly unchanging from one year to the next it didn't much matter if a firm's evaluation procedure was only telling it half the story. Its managers would have become used to the deficiencies and realized – perhaps gradually and subconsciously – how to compensate for them. They might have applied a three-year payback test to an investment in one type of plant, but relaxed it to, say, four years for another. They would have learned that the information the payback period conveys doesn't include everything of relevance to the investment decision, and through experience they would have learned how to make allowances. With new technologies, people do not have the experience to apply such compensations. They need to rely much more on laying out their intentions explicitly – on being able to link technological capabilities with the structure of a firm's operations and with its commercial performance. There is little chance that decisions about new technology based entirely on intuition will prove very convincing, or that the outcomes will be very worthwhile.

The ideas that will be described in the following chapters together provide a way of thinking about the value of technology. They don't go so far as to lay down an exhaustive set of procedures, and they certainly don't say what specific types of system are likely to be worth. As with many disciplines, it is the way of thinking about the issues that is useful and enduring: settled conclusions and specific policies are usually more questionable at the outset, and they tend in any case to be overtaken by events. It is also much more difficult to produce policies that are general enough to be applicable in all types of company than it is to recommend how companies should go about forming their own policies.

Scope

There are two things that set the limits to this discussion. The first is its concern only with the properties of a system that have a material effect on a firm's finances. Issues of high policy, such as those that concern a firm's ethical standing, are ruled out of bounds; as are questions about how technology affects the way people behave, and whether technological finesse is worth pursuing for its own sake. This makes the scope seem quite limited, but in practice these other factors tend to be drawn into the analysis whether they are wanted or not. For example, an investment in new systems sometimes improves morale, and morale has an important influence on financial performance. Morale therefore acts as an intermediate variable, if you like, although it is neither a technological property nor a financial one.

The second boundary is set by restricting the discussion to one about computer-based systems in manufacturing firms. For the most part, it will in fact be just the more distinctive computer applications and the way in which they are interconnected that will be described. This includes, for example, anything that is intended to promote flexibility, quicken the introduction of new products or improve people's motivation. A system whose only effect is to reduce the consumption of stationery is probably too uninteresting to be worth including. Having diffuse and complex types of effect does not, necessarily, make a system more valuable, but it almost certainly makes it harder to value.

Layout

There are several lines of argument that are pursued here. Some of them apply at a greater level of abstraction than others, in the sense that they are concerned with general conditions and long-term trends; they are more useful to the planning process than they are for sanctioning individual developments. Others apply at a more detailed level, and they are likely to be a more useful basis for making decisions about well-defined, short-term issues. The general direction of the discussion is to start out with the more abstract ideas and progress to the more concrete near the end. The former may not be to everyone's taste and they can be skipped if this is not where your interests lie. Doing so, however, runs the risk of studying nuts and bolts in ignorance of the bigger structures.

There are three main parts to the book. The first is about the changing nature of manufacturing companies and how computer technologies are connected with these changes. It is concerned more with changes of kind than with those simply of degree: economies of scope and the boundary between markets and hierarchical organization are two of the main themes. The description of the technology is based on a number of ideas about what makes computer-based systems distinctive, how they influence the basic nature of a firm's processes, and how their financial effects can be thought about in a consistent and comprehensive way. The emphasis at this stage is more on the whole – whole organizations and whole systems – than it is on the constituent elements.

The second part is concerned with the yardsticks by which the value of specific technologies can be gauged. Once a firm has a measure of a development's probable value it can take a number of different kinds of decision about how to proceed: it can choose the most suitable among a number of conflicting courses of action; it can establish a scale of priorities for adopting developments that do not conflict; and it can determine whether any one development is, on balance, likely to add to the value of the firm rather than diminish it.

The third part deals with the practice of making investments in new technology. It looks at ways in which people can gain an understanding of the risks their firms run when they pursue a particular course of action, and it looks at the sequence of activities into which the appraisal typically fits. Several of these activities appear to dilute the impact of an appraisal, because they introduce such issues as personal preference and personal influence. These tend to act in unpredictable ways, and they usually remain tacit (and therefore closed to examination). On the face of it they detract from the simple, testable logic of a financial calculation. On reflection, however, they are often useful in winning commitment and in shaking out bias. They help accommodate uncertainties and conflicts.

Many of the ideas described here are debatable for one reason or another. Sometimes it is for reasons of principle, and other times it is because it is difficult to put them into practice. Where these reservations are especially relevant to the way in which an appraisal is in fact conducted, they will be mentioned. Mostly I shall avoid repeatedly pointing out that a particular idea is one way of looking at things and that there are others which have found favour with different people at different times.

It is worth describing at the outset the point of view on which these ideas are based, and this will be the subject of the next section. It will, I hope, illustrate why no appraisal system will suit all of the decision makers all of the time. But the issues described there aren't essential to a reading of later chapters, and those interested in the application (rather than the origin) of the tools of analysis used later might want to move straight to the next chapter.

1.2 Points of view

All the models used to think about and act upon abstract ideas like economic value are simplifications. If an appraisal were as complicated as the effects of the development to which it is applied it would probably never be finished: there would be too much information to gather and there would be too many calculations to perform. But simplifications plainly have their limits. They are normally made with a particular end in mind, and once a model is used for purposes significantly different from this end it is likely to produce unconvincing or irrelevant results. It would be whimsical to use a model that assumes we know the future with perfect certainty if, in fact, we felt that uncertainty had a material cost.

There are a number of straightforward ways of characterizing these models, and by running through these it will become evident how far we can take the ideas described in later chapters.

Normative and positive

A first distinction is made between models that are intended to prescribe how people *should* take decisions, and models that describe the way in which decisions seem to be taken in practice. Since normative approaches (the former) are applicable to problems that are quite different from those suiting positive approaches (the latter), they need to be judged on different grounds. The main test of a normative model's validity is whether its assumptions are consistent with the beliefs and intentions of the person who is expected to use it. And there are of course a number of other criteria that determine a model's usefulness – its generality, expressiveness and simplicity for instance. The test of a good positive model, on the other hand, is whether observable behaviour is consistent with the model's predictions. It is there to explain how things really happen, and it will be convincing only if it says something correct about a process before the process takes place.

The discussion here is concerned with ways of making decisions about technology in a more informed manner than at present – so it is mainly normative. But it is worth paying attention to ideas about how people make decisions in reality, since the models will not be used if they fail to capture the type of information that decision makers will really act upon.

It is important to get some sort of a feel for how far normative models can be taken. Taken too far they will yield results that will either be ignored or cause people to make questionable decisions. If they are not taken far enough, the decision-making process will be too hazy, too intuitive and perhaps inconsistent from one time to the next. The main factor that limits the models' usefulness is often the assumption that is made about who should be the beneficiary of a particular decision. It is sometimes assumed, for instance, that a firm's managers set out only to increase as much as possible the returns to its shareholders. Most models also assume that decision makers behave in an entirely rational way: that they will take account of every piece of information they have in a way that is consistent with a simple economic objective. Both assumptions are suspect as statements of what happens in practice, but they often suggest an ideal towards which people are prepared to work.

Normative models are often founded on a small number of principles on the basis of which interesting consequences can be derived by purely logical arguments. They therefore tend to be concise and internally consistent. They are inherently suited to decision making by calculation. Unfortunately the real world is complex and unpredictable, and an alternative way of building models is to observe examples of real behaviour and then generalize upon them. The problem with these ideas, however, is that the generalizations they incorporate are not always satisfactory: it is not always obvious when an event is the product of special circumstances and when it is not. Neither is it always obvious which of a number of alternative principles lie behind a specific event.

Maximizing and non-maximizing

A second distinction is drawn between models that are based on maximizing certain quantities (perhaps subject to a number of constraints), and those based simply on raising certain quantities above predetermined thresholds. An obvious example is the contrast between the idea that a firm's managers should maximize its earnings, and the idea that they need only achieve sufficient earnings to maintain a certain dividend. The second approach is probably a much better description of what goes on in practice: people who make decisions generally do so with a number of goals in mind (not just one), they are often uncertain about the effects of different courses of action, and they sometimes have very little time in which to make their decisions. It is sensible, and perhaps essential, in such cases that they set out to satisfy a number of minimal criteria, rather than maximize the effects on only one.

Maximizing models, however, have their uses, and their simplified concentration on a narrow range of issues is sometimes a powerful drive. This simplification is made on the basis that it can pre-empt some of the conflicts that would otherwise cause an organization's work to become directionless and incoherent. It is rarely made in the expectation that everything that goes on in a firm will satisfy the assumptions the model makes. But such models *do* become inappropriate when it is an accurate picture of real behaviour that is needed, or when the quantity that is to be maximized does not represent a central focus for the firm.

I will hedge my bets on the ideas described later. The main yardsticks used for assessing the value of a new development are based on the principle of maximizing a particular quantity, but they can be used perfectly well in more limited ways. You might, for example, predict earnings using a certain model in order to make the choices that are most likely to maximize these earnings. But equally you could use the same model simply to make sure that earnings are likely to be high enough to meet the expectations of a company's financiers.

Certainty and uncertainty

A third distinction is drawn between models which assume that perfect information is available and fully taken account of by decision makers, and those that don't. The latter make explicit allowance for the inevitable uncertainty that surrounds the future, for people's inability to use all the information that is available to them, and occasionally for their characteristic biases and mistakes. In some cases it is feasible to build models that can demonstrate how people act when the uncertainty is a result of having to wait for another decision maker to make the first move. In other cases it has been possible to say what degree of uncertainty should be traded against the costs and delays associated with gathering more relevant information. The idea that it is a set of expectations (as much as firm knowledge) that motivates decision makers has proved to be a good way of understanding a number of phenomena that otherwise appear to be inexplicable.

As uncertainty is such a prominent element of decisions in commercial firms – especially when they concern new technologies and new techniques – the yardsticks we use for appraisal must have a way of dealing with it. The more explicit and natural and practical the uncertainty mechanism is, the more convincing will be the result of applying it. Unfortunately, it will become evident that it is often just as hard to be certain about the degree of uncertainty one faces as it is to be certain about costs, benefits and so on. So a part of the discussion will inevitably be about qualitative reasoning: about the processes of judgement we can use when significant elements of the decision to invest in a new technology cannot be compared with one another in a simple, arithmetical way. Another part will be about how uncertainty arises, and the manner in which bias usually becomes evident.

Deterministic and non-deterministic

A final distinction that is sometimes convenient is the one drawn between determinism and non-determinism. A deterministic process always produces the same output whenever it is presented with the same input, while a non-deterministic process produces outputs with a degree of randomness (even if it is slight). The best examples of non-deterministic processes are performed by people – because people have such things as memory and mood, and their reasoning is often extremely difficult to fathom. The fact that they have memory means that they store knowledge from one event to another, so that they frequently modify their response to successive events of the same kind. The fact that they have mood means that both their perception of the outside world, and the pattern of values they apply to it, change from one day to the next in an apparently random way. Whether we think we can know exactly how organizational processes take effect occasionally makes a big difference to the line of argument we pursue.

An example of a deterministic model is a firm's production function. This simply plots the relationship between the flows of its inputs and outputs; it records the maximum level of output that can be obtained by using given levels of each type of input. Typically, one of the inputs is a measure of the aggregate labour a firm employs, and another is a measure of the capital it uses. A particular production function really only applies for a specific level of technological sophistication, and the adoption of a new technology will normally be expected to shift the function in some way. The most important aspect of this model is that it assumes that the production function is beyond the control of the company: that while it uses a particular technology there is a known, pre-defined limit to the relationship between its inputs and outputs.

The application of production functions can become quite sophisticated. There have, for example, been proposals to use an index of information technology as one of the inputs.[1] This index would place the extent of the technology against a numeric scale: perhaps the number of automated teller machines in a bank, or point-of-sale devices in a retailing organization. In principle, the firm's managers could work out the resources they needed to produce a given level of output, and how they could trade more of one type of resource for less of another. They could discover, for instance, how much labour might be shed when a new element of information technology is introduced.

Unfortunately, production functions are rarely a convenient model to apply to the real world. The distinction they draw between changes in the mix of inputs and shifts due to technology is usually a meaningless one – firms rarely change the amount of capital or labour they use without making some sort of technological advance. Production functions are very difficult to measure: a firm will know its current output and the level of its current inputs, but it is unlikely to know with much accuracy what will happen when it makes significant adjustments to these. The most serious problem, however, is that production functions imply a deterministic view of non-deterministic processes. In a real factory there is a vast number of factors that determine how much output can be produced from given levels of inputs, and elements such as morale, autonomy and cohesiveness make big differences. It would be wrong to think that we could ever put these into a giant equation which we would solve in order to plot the future course of a business's development.

For the most part, I therefore want to avoid deterministic models of the manufacturing process, and especially the idea that the effects of new technology can be understood by constructing production functions. They are sometimes useful for explaining certain situations, but rarely so for making quantitative predictions about the value of specific developments. (They are intended in fact to help with analysis at more aggregated levels.) We will, of course, be interested in the way that technology affects the relationship between input and outputs, but we cannot assume that we shall ever be able to express this in terms of simple arithmetic formulae. Neither can we assume that there are natural, unvarying limits to what can be done with a particular technology.

Notes and references

1 Alphar, P. and Kim, M. A microeconomic approach to the measurement of information technology value. *Journal of Management Information Systems*, **7**(2), 55–69 (1990)

Part I The Background to New Systems

2 The changing nature of manufacturing

When our first parents were driven out of Paradise, Adam is believed to have remarked to Eve: 'My dear, we live in an age of transition'.

W. R. Inge *Assessments and Anticipations*

A number of general themes have become evident in most markets for manufactured goods and in many of the industries that supply them. Although our main concern will be with the value of specific technologies, these themes help to explain the circumstances in which individual systems are put to work, and they provide a distinctive thread tying one investment to another. Most of the subsequent chapters are about the detailed analysis of opportunities that have, somehow, already been identified. By looking at the more wide-ranging ideas of how manufacturing firms are developing we have a consistent way of searching for such opportunities – of choosing developments that complement rather than offset one another in their effects.

It makes sense to begin by looking at the changes taking place in a manufacturing firm's environment, and to follow this by discussing the structures that a firm can introduce to cope with these changes. So the first part of this chapter discusses the way in which the demand for particular goods typically develops – the stages it goes through and the manner in which the transitions between the different stages take place. The subsequent parts look at the supply side: at issues that concern the organization of the manufacturing process. The intention is to examine the particular ideas that will match most current and most imminent stages in the development of changing demands.

This discussion is not a comprehensive one, in the sense that it doesn't examine all facets of the manufacturing organization. It is concerned only with changes, with significant departures from past practices and with issues that are bound up with the application of new technologies.

2.1 New patterns of demand

A convenient model with which to consider new patterns of demand is the sequence of market development proposed by Bolwijn and Kumpe[1]. Their suggestion is that the focus of a market's attention evolves in outline through a series of phases. The first phase is one in which a product's price is the main pivot of competition – making a producer's efficiency the obvious quantity by which his performance may be judged. In other words the emphasis is placed on minimizing the consumption of resources. For many markets this phase persisted through the 1960s.

The second phase is one in which a product's quality or fitness for its intended purpose is given most prominence. This suggests that the accuracy and integrity of the producer's processes and organization is the outstanding determinant of his performance. This is supposed to have been associated with manufacturing in the 1970s. Quality is a measure of the certainty people feel about the prospects that a product meets its specification, whether this is advertised or whether it is tacit. It isn't simply traded against other product attributes (such as technical performance); so from the start of the second phase a producer who cannot provide it loses the goodwill of his customers.

The third phase centres on the range or variety of products available to meet a certain need, and now the performance of suppliers is primarily determined by the flexibility of their operations. It is an element that was perhaps most noticeable in the 1980s.

The fourth phase is one in which the market wants products that are unique and innovative. This need to innovate will perhaps be the dominant feature of manufacturing in the 1990s. The tendency, however, is not that earlier requirements will be displaced by later ones, but that they will be joined by them. So in practice many of today's companies compete simultaneously on price, on quality and on variety.

A particular form of organization is often associated with each of the four phases just listed. Efficient firms, competing by price, tend to embody a high degree of specialization, exerting administrative control through fairly rigid hierarchies. Firms competing on quality, however, are more likely to stress effective mechanisms for communication and co-operation. To some extent, this means dismantling some of the hierarchy found in the firm in its earlier form: there certainly has to be a good deal of horizontal dialogue, as it were, to reach even the rather mediocre levels of procedural quality demanded by recent standards. In the third phase, a flexible firm needs to decentralize controls and integrate the activities taking place at different points in the manufacturing process. It needs quick changeovers, for which decision making is best done at as low a level as possible. It needs quickly propagated instructions and quickly gathered measurements to keep the right products in production. The innovative firm will, in addition, have to allow people a larger measure of participation than most currently enjoy. Such a firm will have little prospect of success if its administrators stifle relevant and practical ideas – or if they fail to create the conditions in which ideas can surface in the first place.

Because the transition to a new phase relies on the capabilities associated with previous ones, it seems unlikely that any of these phases could be bypassed. To capitalize on flexibility, for instance, would be almost impossible for a company that didn't already have processes and products of high quality. Fast product changeover isn't feasible if the transition from design to production is muddled, inaccurate or untraceable.

A transition between phases itself commonly passes through a number of obvious

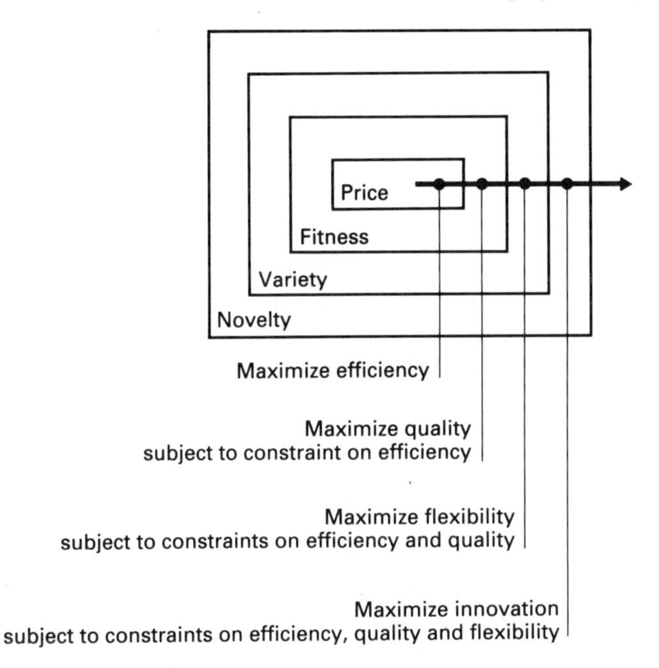

Figure 2.1 Typical demand development

stages. During the first stage the new pattern of market demand is simply ignored or denied by a company's managers. In the second it is seen as a problem – an external event to which a company has to respond. It is a time when the actions of competitors will be watched closely. In the third stage the new demands are seen as a source of opportunity, and a way of gaining some sort of lead over those competitors that are still dithering. Whether the passage from one stage to the next is very distinct will no doubt depend on a firm's style. If its planning processes are explicit, and if it tends to make progress by consensus, the transition is more likely to have well-defined stages. In other cases we might expect individual managers to make the transitions in their own understanding at different times, and perhaps in such a gradual way that they don't have to admit to changing their minds.

It is evident from this model that a firm shouldn't be seeking reductions in its cost base alone unless the market remains in its first phase – the one in which sustained competition is based on low prices. If the most significant element in the current market is product range then it makes sense to look for machinery and systems that embody flexibility. When a firm in this phase chooses between one new system and another, it might in fact insist on minimum levels of productivity and accuracy in the processes the systems perform, and then decide between the candidates on the basis of their flexibility – choosing whichever maximizes the flexibility the firm can exploit with its existing resources. In other words, the firm can attempt to raise the level of the key competitive element as much as possible once it has brought other elements above certain thresholds. Figure 2.1 summarizes this idea.

Whether or not this kind of decision rule is appropriate in specific instances, it is reasonable to suppose that the patterns of demand in a firm's product markets will have a considerable bearing on the properties it will expect its internal systems to exhibit.

Some markets in later phases might revert temporarily to a condition in which price becomes the predominant element – during a recession perhaps. But any decisions about technological infrastructure plainly need to be taken with a view to the circumstances that will be experienced in the coming years rather than the following few months. It may be that a market is in transition from one phase to another, or it may be that it is obviously drawing to the end of a particular phase. In conditions such as these it makes sense to plan for the phase that will succeed it, and, at the very least, to attempt to identify when firm decisions will need to be made about the appropriate infrastructures.

The idea that the transitions between distinct phases go through periods of rejection, grudging acknowledgement, and finally a reasonable and active optimism, suggests that firms will experience a series of reactions before properly coming to terms with new demands. Perhaps each of these stages *has* to be negotiated if new developments are to have firm foundations. The act of successfully managing change becomes one in which the organization is taken through the earlier stages with as little pain as possible. Managers and technologists should have the maturity, you might say, to recognize that the transition will most likely prove fruitful so long as progress can be sustained in the face of initial scepticism. In particular, the failure of an individual investment during a transition shouldn't be seen to invalidate the general direction in which change is taking place.

The model also proposes that particular patterns of demand are naturally linked with the structures and the cultures of the organizations that try to meet this demand. This obviously means that the technology isn't enough on its own; it has to be accompanied by changes in such factors as attitude, procedure and organization. It is normal to divide new developments into distinct issues – typically separating the technical from the organizational – in order to keep the problems to a manageable scale. But one shouldn't forget to put the pieces back together again, and there will, of course, be problems in their assembly as there are in the construction of the individual pieces.

It is probably fair to say that at the moment the most interesting and material stages of the demand model are the third and the fourth. The internal developments that will be considered in the next two sections are therefore aimed above all at creating the conditions associated with these two stages. They are meant to help firms increase their flexibility and their ability to innovate while at least preserving the earlier efficiency and quality of their operations.

In some respects, these developments are unfamiliar and occasionally they run counter to intuition. It is worth bearing in mind that in many markets and industries they are only just beginning to become relevant. But this is, of course, the point at which it is best to start thinking about them and planning for them. Otherwise the influence of competition can quickly force a firm into reacting to rather than shaping the events in which it participates.

2.2 Economies of scope

The first of these ideas is based on a principle known as economies of scope. Since it is based on product variety and production flexibility it ought to be especially suited to meeting the needs of the third demand phase. Before looking at this, however, it is worth considering an older principle that has prevailed up to now, and which – in part at least – is beginning to lose its validity: the principle that industrial firms can enjoy economies of scale.

Economies of scale

The presence of economies of scale in manufacturing has been used by managers in their search for greater efficiencies, and it has been used by researchers to explain the structure of markets and industries. It provides a rationale for what seems to be a natural tendency for firms to grow in scale: to seek bigger market shares, to buy more capital assets and to employ more people. It is a convenient argument for industrialists intent on extending their sphere of control.

The basic idea is that the average unit cost of production falls with increasing capacity. It is a long-run effect in the sense that changes in capacity, or scale, reflect changes in long-lived capital plant as much as changes in commodities whose stocks and flows can be varied over shorter intervals.

There are several potential sources of such economies in a firm's manufacturing operations. The first arises because large firms can use specialized labour. Since these firms earn substantial revenues, they can employ large numbers of people, and these people can be allocated highly specific jobs. In such jobs, so the argument goes, they can develop and apply a level of knowledge or dexterity that wouldn't otherwise be possible.

A further source of scale economies stems from the way in which the cost of some forms of capital plant is supposed to increase at a lesser rate than the output it can produce. There is a rule that associates the cost of plant (such as boilers and storage vessels) with its surface area, and the potential output with its volume. And since the ratio of volume to surface area increases with size, the unit costs of the product attributable to the plant decrease with the scale of the output.

Finally, it is sometimes the case that particular inputs to a production process are relatively indivisible – that they can only be bought or made in specific, discrete sizes. These inputs will be wasted to the extent that the amount used for production falls between the available sizes. If the volume of production is low, the level of waste attributable to each unit of output will be high. If the volume is high, the unit level of waste (and hence the unit cost of a product) is likely to be lower.

There are, however, a number of problems with these arguments.They are discussed at length in a survey by Gold,[2] which I shall draw on here. One problem is that the stricter definition of scale economies stipulates that when you test for them you have to keep both the relative proportions of the inputs constant, and the technology of production unchanged. Changing a plant's capacity without changing its technology is very unlikely to happen in practice. This makes it hard for researchers to test for scale economies, and it renders the concept less than useful to managers planning their factories. The looser approach to defining scale economies is simply to look at the relationship between the costs per unit of output and an aggregated measure of the inputs, regardless of whether there are any accompanying changes to variables other than scale. This definition is more useful when trying to prove that economies exist, but it doesn't help to explain their causes. In particular it cannot be used to determine whether lower costs are *caused* by scale: it can only say whether they are correlated with it.

A second difficulty is that the economies of capital plant that stem from the area-to-volume relationship have a very limited applicability. It may be true that an element of a plant's cost is related to its surface area, but the relative importance of that element isn't necessarily very great. Increasingly, the price of control systems predominates over the price of the types of mechanical equipment that demonstrate these economies. This source may remain relevant for certain types of plant used in process manufacturing, but it doesn't seem plausible for the production of discrete goods.

The idea that economies come from specialization is also open to question. It is perhaps only valid in very small firms. Even if a medium-sized firm were suddenly to double in size there might be nobody, as a result, who ended up performing a more specialized job. A more likely consequence would be that services previously bought from specialized outsiders were instead provided by new internal departments. Anyone in fact called on to do a more specialized job could well find the associated repetition and lack of variety dispiriting enough to offset any theoretical savings. And it is commonly the case that wages are higher in larger factories than they are in smaller ones – perhaps by a big enough margin to yield *dis*economies of scale.

There are other instances in which there seems to be a clear association between scale and reduced costs, but it isn't necessarily the case that one causes the other. Sometimes, such factors as the standardization of components and materials are associated with big factories, and one gets the impression that the savings a firm can make are scale economies. But they are not – they are standardization economies. Since it is conceivable that many small manufacturers could make use of standardized components from common suppliers, large factories are not a necessary condition for standardization. Perhaps there are genuine scale economies in activities like research and development associated with the ability of large firms to diversify away risks that they cannot insure against in any other manner. But even this effect is more a function of a firm's development portfolio, as it were, than of its size.

To pursue scale economies is an especially questionable goal in factories that are introducing both flexible manufacturing systems and high levels of integration between their different activities.[3] Here, batches are deliberately kept small to make lead-times short and product introduction fast. The emphasis is on being responsive rather than being big. And, in automated systems, much of the expertise of specialists in disciplines such as scheduling and process diagnosis is embodied in software – software that is almost as readily affordable by small factories as large ones. In other words, the need for large-scale operations is becoming less apparent, and the desirability of small-scale operations is becoming more so.

Economies of scope

The feeling that, with the newer technologies, firms can make much broader use of know-how and capacity suggests that there may be another source of economies. Far from being concerned with the *scale* of a firm's operations, they stem from its ability to cope with variety, or *scope*. The greater the scope of a factory, the more economies it can gain from re-using and re-applying its intellectual and physical resources to the different types of product it produces.

Suppose that a firm were faced with a demand for a given number of product lines. It might decide that they should be manufactured separately – that there should be no common activities during their design, planning and production. It might, on the other hand, choose, to produce them in such a way that a part of the knowledge or material used in the manufacture of one could be applied to the manufacture of the other. In this second case the firm would be attempting to capture scope economies – to find resources sharable among its outputs.

There will be limits to such economies, and these are probably going to be set by the problems that a large amount of variety causes, such as the chaos that is eventually associated with doing many different things at once. It will, of course, be a matter of some interest as to where in practical terms these limits lie. It will also be of interest to know just what we mean by *variety* (or scope) and whether it is something that will really help people understand the best way to organize their factories and make use of new technologies.

That there are economies associated with variety is in some ways rather surprising. It might be more natural to suppose that coping with variety has a net *cost* rather than a gain, because it appears to be more expensive to use general-purpose systems than to use highly specific ones. The sophisticated control gear and handling equipment needed for a flexible manufacturing system (FMS), for example, adds considerably to the cost of the machine tools it contains. When you consider all the additional work needed to graft the FMS on to a complex organization, the extra cost is far greater. Surely it would be easier to build specialized plants for each product line rather than attempt to produce them with common resources? The argument about scope economies essentially suggests, however, that the costs of generalized, flexible systems are outweighed by the benefits of sharing resources and exploiting them more intensively.

Of course, if a company is unable to market a wide variety of products this argument breaks down, But it seems that customers *do* want variety – they want products that match their particular needs and circumstances. In the third phase of the demand model described in the previous section, companies compete by being flexible: by offering options. In speaking of the internal efficiencies of producing a variety of products, with which economies of scope are essentially concerned, we shouldn't therefore be drifting too far from what is commercially desirable. By finding scope economies one can make a virtue out of a necessity.

In any case, much of this variety is only potential: at any given time a firm may be unaware of the precise nature of the variants it will be called on to produce. By employing flexible systems, the firm is more able to respond to changes in this variety over time.

A proper definition

It is as well to adopt a thorough definition at this stage: this will make it easier to be precise about what scope economies are and how they can arise. Although it will rarely prove necessary to speak about these economies on a formal basis (and although you might want to skip this sub-section if you find definitions irksome), a more precise understanding is always a useful fallback when informal lines of reasoning seem vague or ambiguous. I shall mainly follow the notation used in a paper by Panzar and Willig.[4]

The definition of scope economies uses the idea of a firm's *cost function.* This function records, in theory, the minimum cost of producing a particular product, given the price of the inputs used in the production process. It is not explicitly a function of the input quantities because it applies only to the most efficient combination of inputs needed to produce the stated output. For various reasons it is not a very useful model for managers struggling with the common uncertainties of a manufacturing business, but it is a helpful stepping-stone to understanding the influences on a firm's performance. Here, we are interested in the cost function associated with a particular *set* of products. If we call the list of these products in specific quantities y, then the function that tells us the minimum cost of producing them will be called $C(y)$. Strictly speaking, this cost function also depends on the prices of the inputs, but as these are going to be held constant they will remain hidden in the definition.

Suppose then that a firm can produce a set S of different products in quantities $y(S)$. S might, for instance, be the set {Product 1, Product 2, Product 3} and $y(S)$ might be the list [1000, 2000, 1500], denoting the quantities of each of the three products in order. Now think of all the ways of dividing up the production of S among separate factories. (In fact these might just be separate areas of the same physical factory: the essential point is their separation.) We can then speak about a set T of smaller sets T_i, each of which represents the output of one of these factories. Since there are few products, the possibilities are very limited; they are shown in Figure 2.2. The partitions shown in the figure can be written thus:

Partition 1: $T = \{T_1, T_2\}$ = {{Product 1, Product 2}, {Product 3}}

Partition 2: $T = \{T_1, T_2\}$ = {Product 1, {Product 2, Product 3}}

Partition 3: $T = \{T_1, T_2, T_3\}$ = {{Product 1}, {Product 2}, {Product 3}}

Partition 4: $T = \{T_1\}$ = {{Product 1, Product 2, Product 3}}

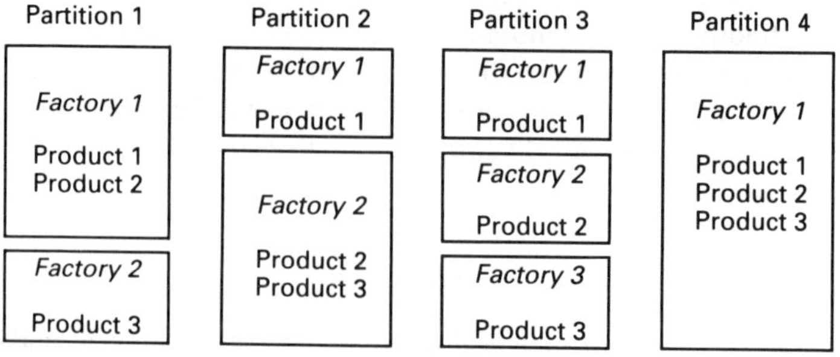

Figure 2.2 Production partitions

Note that, for a specific partition, its members are produced with *no* common inputs. In Partition 1, for instance, Product 1 and Product 2 are produced together, but Product 3 is produced quite separately. Our test of whether there are economies of scope will be to compare the cost associated with producing a partition of S with the cost of producing S itself. In other words we will be comparing the cost of production in separate, more specialized factories with that of production in just one factory of wider scope.

The cost function for producing a small set of products T_i, in a partition T, is $C(y(T_i))$ and that of producing the products all together is $C(y(S))$. Then, as a matter of definition, there are economies of scope with respect to the *particular* partition T if

$$\sum_i C(y(T_i)) > C(y(S))$$

In the example of our three-product firm, it might be the case that, in Partition 1,

$$C([1000 \text{ of Product } 1, 2000 \text{ of Product } 2]) + C([1500 \text{ of Product } 3]) <$$

$$C([1000 \text{ of Product } 1, 2000 \text{ of Product } 2, 1500 \text{ of Product } 3])$$

In other words, the cost of producing all three products together is less than that of separating out the production of Product 3. Perhaps a production system has to be duplicated when Product 3 is manufactured elsewhere. There are said to be *dis*economies if the inequality operates the other way: that is, if

$$\sum_i C(y(T_i)) < C(y(S))$$

These tests make it clear that a scope economy does *not* arise because a given factory has lower costs when it produces many types of product than when it produces just one: such an effect is of course very unlikely. It is, instead, a situation in which costs are reduced by producing in one factory many types of product that would otherwise be produced in several factories (or at least, if in one factory, with no common processes).

These inequalities won't be used to derive any further results, although you might care to look at Panzar and Willig's paper to see the implications they have for certain issues within the domain of industrial economics.[4] The important thing to bear in mind is that one can meet a demand for product variety in different ways: either by producing the different products with specialized, separated resources, or by producing them with flexible, sharable resources. If we can find resources that are sufficiently sharable, we can gain economies of scope.

Some sources of scope economies

What, in reasonably concrete terms, are the main sources of scope economies? What sorts of technology should we expect to help us obtain them?

An article by Talaysum *et al.* suggests a number of possibilities.[5] The first is machinery or plant of some sort that is flexible enough to be useful in the production of more than one product type. An unexpectedly high demand for one product can be met by using any capacity released by an unexpectedly low demand for another. This, on the face of it, seems to be rather a marginal effect. It is based only on differences between the volume of products forecast when the plant was introduced, and the demand that materializes at any subsequent time. In practice, this disparity can be very wide over short periods of time, even if over a longer period the level of demand exactly matches original expectations. Given the uncertainties of predicting market demand several years in advance, it is perhaps rare that such expectations are actually met – even in the long run.

The type of plant that most obviously lets factories switch capacity between one product line and another is the flexible manufacturing system. FMSs usually incorporate machines adaptable to the production of quite distinct product geometries, sizes, assembly structures and so forth. Above all, the FMS as a whole offers such quick and cheap mechanisms for reconfiguration that it becomes attractive to change between product lines at short notice. Machine tools, robots and transport equipment can, in principle, be quickly reprogrammed, and schedules within the system recalculated to reflect the impact of changes in the order of production. It has perhaps been a common experience that such potential capabilities are not fully exploited for one reason or another. The inability of many firms to get information such as part programs and schedules flowing freely enough throughout their organizations is a prominent cause. But flexibility is at least now becoming a property of the technology, if not of the organization that puts it to use.

A second important source of scope economies lies in the ability, with information technology, to make knowledge and expertise much more a kind of *public good*. A public good is something of material usefulness with two essential properties: non-rivalry (in that its consumption by one person doesn't reduce the quantity available to others) and non-excludability (the fact that people who don't explicitly pay for it are not prevented from consuming the good). Precisely because a public good is widely exploitable and not depleted by consumption, it is sharable: because it is sharable it can offer economies of scope.

Consider, for instance, the possibility of capturing in a knowledge-based system one man's know-how about repairing machine tools. Once his know-how has been encoded in software it can be copied a limitless number of times at virtually no cost. Once copied it can be replayed on a similar computer elsewhere in the factory. If it is made available across a network, it is accessible to anyone who wants to make use of this man's knowledge. And as it is never worn out or reduced in value when it is applied, it has non-rivalry in consumption. It is, in other words, a quasi-public good – public, at least, within the company. And it therefore becomes a sharable input.

Knowledge-based systems are, of course, just a particularly good example of this phenomenon. It is the fact that information is not destroyed by its application that makes information systems of any sort non-rivalrous in consumption. It is the fact that software is so easily duplicated and so easily invoked across networks, that makes information technology non-excludable. The information doesn't have to be a particularly arcane article of knowledge, nor the technology a particularly sophisticated piece of artificial intelligence, for the combination to offer scope economies.

There are two other main sources of scope economy that are worth mentioning, although they are arguably less prominent than those just discussed. The first is found in cases where the production of one product line produces waste material that could, in theory, be used in the production of another. Indivisibilities in the quantities and shapes of incoming materials mean that they cannot be purchased exactly to match the needs of single product lines individually. At normal production speeds, without significant computing power, the optimization of material usage has often been impossible: firms have in such cases been unable to share the material between product lines. It needs complex algorithms and it needs a good deal of information about the geometric properties of the product. This information may be available, in an implicit form, in part drawings – but it has still to be re-cast in a format that is suited to calculation. With nesting software, linked to computer-aided design (CAD) systems, this can in many cases now be done much more readily.

A similar line of reasoning can be applied to people's expertise. Since many in the factory work increasingly on indirect tasks, their efforts are distributed over more than one type of product. Whenever their work on one product type does not exclude working on another, this labour is a shared input and has economies of scope. This capability is not simply a result of new technologies, but it is a trend which they can promote. Systems which capture general rules applicable to many kinds of product or process contribute to scope economies in this way. One might contrast these with systems that can only record highly specific types of information. Thus expert systems are valuable not only because they enable firms to disseminate knowledge, but also because they enable them to capture knowledge in relatively general and widely applicable forms. This is perhaps especially the case with engineering systems – those that capture the principles of efficient, testable and producible design.

In some ways, variety is a clear counterpart to standardization. The principle that customers increasingly want products tailored to their highly specific needs, and the principle that there are economies in doing so, suggests that standardization is a strategy both ineffective for marketing products and inefficient for production. This is really only the case with a firm's final outputs, however. The idea that scope economies are associated with sharable inputs in fact suggests that *within* a factory the standardization of components and methods would be an ideal well worth pursuing. The factory can then economize on the expertise and labour needed to engineer a component re-used in many product lines. It is left to the point at which final products are put together to adapt them to the needs of particular customers.

Finally, it is possible that there are scope economies associated with financing. This hinges on the idea that people and institutions providing funds demand a premium for risk – the possibility that a firm's earnings will turn out to differ widely from those forecast over a particular period of time. The greater the risk, the greater the premium and the greater the firm's financing costs. This issue will be considered in much more detail in the later chapter on the present value yardstick, where it turns out that the premium is *not* directly related to such a simple idea of risk. However, at this stage we can suggest that the broader the variety of product lines a firm produces, the less sensitive (all other things being equal) are its earnings to economic factors beyond its control. This means less risk and smaller funding costs.

In practice, the rider that all other things must be equal makes a big difference: companies producing a very wide range of products can spread themselves too thinly, in the sense that they find it increasingly difficult to bring managerial talents to bear on diversified businesses. This suggests that, if there are scope economies associated with financing, the point at which these turn the corner to become diseconomies is within all too easy a reach.

Flexibility for scope

Economies of scope are attached to the *variety* of outputs that a factory produces. In practice, variety is a difficult concept to work with, and it would be nice to use something simpler. What constitutes the same product line, and what is something clearly differentiated from it, will not always be clear – particularly when a firm produces many minor variants of a general product type. In engineering firms that work to contract (in contrast, say, to mass producers) every piece of work may in some respect be different. It is hard to attach a measure of variety to this type of product. Perhaps most problematic of all is the uncertainty about what the firm might do in the future. In principle, if it wants to plan its infrastructure and processes to take advantage of scope economies, it would have to predict for many years ahead the precise variety of products it expects to produce. So while it continues to rely on the concept of variety, the idea of scope economies has little operational worth: you would be hard put to use it as a tool to guide industrial decision making.

The article by Talaysum *et al.* suggests that wherever the arguments about scope economies refer to variety,[3] a similar line of reasoning can be applied to *flexibility*. Provided that a firm exploits it by producing a variety of product lines, one can speak about flexibility as being the re-application of common resources to different types of output. In other words, flexibility yields scope economies. Typically the common resources are forms of physical production plant (such as FMSs), but they might equally be forms of knowledge (such as design or diagnostic know-how). It therefore makes sense to substitute flexibility for variety in the remainder of the discussion. This will be easier to plan for, and it fits quite well with what ought to be one of the main characteristics of a plant using advanced technology. Using economies of scope, a fairly direct connection can then be made between flexibility and the efficiency of a company's operations: such economies provide a rationale, if you like, for introducing flexible practices and flexible technology.

What we can then say is that, up to a point, the greater the flexibility of the factory, the lower the unit cost of its products. So when making long-run decisions – decisions in which any part of the firm's infrastructure may be changed – the factory's managers face a curve that might look something like the sketch in Figure 2.3.

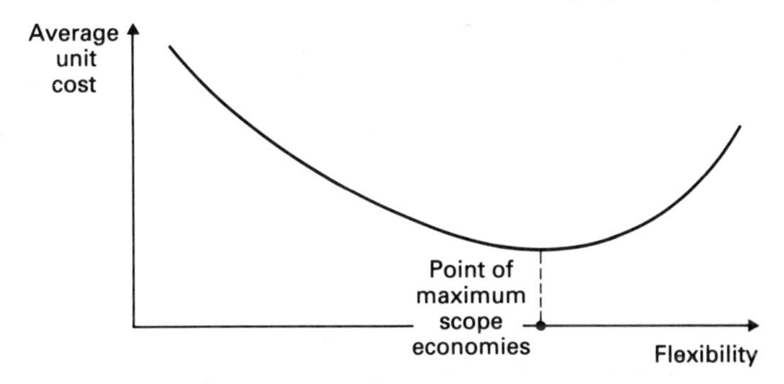

Figure 2.3 Economies of scope

That such a curve can be drawn at all is a result of substituting flexibility, a single dimension, for variety, a thing of many dimensions (each possible product line is a distinct dimension). Faced with a curve of this type, the objective of the firm in the long run would sensibly be to increase flexibility to the point at which the marginal return to this flexibility is zero – to keep buying flexibility until the extra benefit only just balances the extra cost. This point lies at the minimum of the unit cost versus flexibility curve.

In practice, it is of course very difficult to establish when a firm has reached its optimum degree of flexibility. Probably no-one would attempt to draw a real curve corresponding to the sketch just given. Even if a firm went to the considerable trouble of experimenting on itself in order to obtain a number of points on the curve, it would find it very hard to measure the results. There is no well-established, quantitative measure of flexibility (although some interesting attempts have been made in the academic literature), and many companies have little idea of the costs genuinely attributable to specific products. So there is little else to do but exercise a degree of informal judgement: to say, perhaps, that it seems worthwhile investing in a system that will produce products *A* and *B*, but to draw the line at a system that will also make products of type *C*. Although we cannot, it seems, make quantitative predictions we do at least have a clear line of qualitative reasoning. If a manager can keep in his mind's eye the idea of economies of scope while deciding on how flexible his operations should be, he will make decisions that are at least defensible.

It is an interesting fact that the average flexible plant in the USA last year turned out 10 different part types with a volume of some 1727 units each: the average Japanese flexible plant turned out 93 types with a volume of 258 units. Moreover, the Japanese introduced 22 brand new types for every one introduced by the Americans.[6] There is no reason to suppose that the Japanese have taken flexibility too far, so the Americans can, at least in aggregate, assume that they are somewhere on the left-hand part of the scope economies curve.

Types of flexibility

To round off this section it is appropriate to look briefly at the nature of flexibility. This ought to make it apparent how flexibility can be obtained, and how the flexibility of different systems can be compared.

An article by Slack[7] suggests that there are two basic determinants of flexibility – a system's *range* and its *response*. Range refers to a system's ability to enter different states, and it will vary from time to time depending on the state the system is actually in. For example, an FMS with a number of raw materials on input pallets will probably have several possible next steps. It could perhaps transport any of the materials to any of the machine tools or other process stations within the system. The greater the number of these options the greater the system's flexibility. Response refers to the ease with which the system can move between different states. The advantage of an FMS typically lies as much in the speed at which it can change a product's process route as in the number of process routes that can conceivably be followed within it. Any system is ultimately flexible given the time and resources to transform its structure to suit a new application, so the ease with which new states can be adopted is an essential part of the definition of a flexible system.

One can also consider the distinction between outward-looking and inward-looking types of flexibility. The first is something that allows firms to react to outside factors that it can't control – such as economic conditions, the state of its markets and feelings of its customers. The second is concerned with the ability of the firm's systems to react well to uncertainties and surprises within its boundaries – process failures, for instance. The first type tends to be the more prominent: it is concerned with giving the factory's managers a number of options for less frequent decisions of a larger impact than those allowed by the second type.

Finally, one can reverse the logic of the previous subsection and look at flexibility in terms of its contribution to scope economies. Machinery that lets a firm produce markedly different products scores more highly on this scale than machinery that lets a firm produce the same products by different methods or routes. Having many instances of the same type of process plant, for instance, gives a firm a certain amount of flexibility for coping with machine breakdowns, but not for extending the variety of its products.

Rigorously measuring flexibility is a much harder thing to do, and there is probably no simple approach that has yet achieved any measure of acceptance in industrial firms. You can get an idea of the lines people are thinking along from papers in the operational research literature. A paper by Kumar, for instance, describes the use of information theory to model the flexibility of an FMS.[8] The FMS is represented by the collection of possible states it can adopt, and by a transformation matrix that shows how likely it is that the system will proceed from one state to another.

Suppose that such a system has at any point n options of what to do next, and that there are probabilities p_i that the ith option will be adopted. The flexibility yardstick should then be a function of all the p_i and it should demonstrate a number of properties:

- It should register the greatest flexibility when the system can adopt any state with the same likelihood; that is, when the probabilities are all equal: $p_1 = p_2 = \ldots = p_n$. (These ought in fact to equal $1/n$ since the probabilities must add up to unity.)
- This maximum flexibility should increase as the number of possible states, n, increases.

- The flexibility function should be lowest when the system can adopt only one state; that is, when there is a j such that $p_j = 1$ and $p_i = 0$ (for all $i \neq j$).
- The function measuring flexibility shouldn't change if a new option is added with a negligible probability of being adopted.

These, and some slightly more technical requirements, fit nicely with a quantity from information theory known as *entropy*. This, as a matter of passing interest, is calculated as

$$S(p_1, p_2, \ldots, p_n) = -\sum_i p_i \log_2(p_i)$$

For example, if a system could adopt one of four possible configurations with equal likelihood, then the entropy is

$$-[0.25 \times \log_2(0.25) + 0.25 \times \log_2(0.25) + 0.25$$

$$\times \log_2(0.25) + 0.25 \times \log_2(0.25)]$$

$$= 2$$

If it were more flexible and could adopt eight configurations then the entropy would be

$$-[0.125 \times \log_2(0.125)] \times 8$$

$$= 3$$

If, however, it were so inflexible that it could adopt only one configuration then the entropy would now be

$$1.0 \times \log_2(1.0)$$

$$= 0$$

This is not however the end of the matter because the entropy of a system depends on the particular state it is in when the measurement is taken. If the system were an FMS, each movement of a workpiece, a robot or an AGV would mean that the FMS took a different degree of flexibility. To produce a constant measure for the system as a whole, Kumar[8] has to consider how it would reach a steady state. I shan't pursue the argument any further as it becomes quite involved – but it is important to note that this approach is restricted to working with the *probabilities* of new states being adopted. It says nothing about how *worthwhile* any of these states is, relative to other possible states, and it is therefore somewhat restricted in the information it conveys. The approach also concentrates on Slack's range element,[7] largely ignoring the question of response.

The model does, however, lead to some interesting results. Among them is the fact that, in general, the conditions under which productivity is maximized are different from those under which flexibility is maximized. One cannot, it seems, have the best of both worlds. A different approach to the same problem, based more on finance theory, is described in a paper by Kulatilaka: this provides an interesting contrast for those wishing to explore the subject further.[9]

The discussion of scope economies (and the definition provided a little earlier) was based on the assumption that the services sharable between one product line and another were produced internally – by the same firm that uses them. It assumed, for instance, that the maintenance expertise captured in an expert system was that of an employee in the firm that operated the relevant machines. This, of course, needn't be the case, especially now that mechanisms such as expert systems make it easier in principle to exchange expertise between different companies. In other words, the sharable services might be bought in a market from another firm altogether, instead of being generated internally.

The same is true of flexible manufacturing: if contracting companies can combine the ability to carry out production processes of widely varying types with the ability to achieve high throughput levels, there is much less need for the principal manufacturer to introduce FMSs to his own factory. In this case, economies of scope – and the point of having one firm produce a variety of product lines – disappear. If the services in question can be bought and sold in a market to everyone's satisfaction, then the effect of scope economies is pushed back a level to suppliers of the shared services.[10] The suppliers may well then decide to share the economies with their clients under the influence of competition with other suppliers.

Exactly what determines the way the line is drawn between exchanges conducted in a market and those conducted within a single firm, is the subject of the next section. Our main motivation, however, in looking at this division between markets and hierarchies will be to explore ways of meeting the conditions suggested by the fourth phase of the demand model – conditions that promote a firm's ability to innovate.

2.3 Markets and hierarchies

The choice between markets and hierarchies (that is, command structures) as the more appropriate way of allocating materials and services to different activities is an important and practical one. Central planning within large economic systems has been thoroughly discredited because, in practice, planners cannot collect and process information quickly and accurately enough. They have failed to predict consumers' preferences and they have failed to understand producers's motivations. Markets, on the other hand, have their own problems. They have sometimes failed to allocate resources to highly-valued activities, and they have occasionally ignored the unpleasant side-effects of industrial progress.

Central planning often appears arbitrary: it stifles initiative, it is unresponsive to pressing needs for basic resources, and it embodies the idea that a chosen few dictate the preferences and achievements of many. Markets, of course, stir competitive feelings, and give those participants who keep their heads above water a feeling of control and independence. Yet competition is sometimes ruinous and it wastes effort on promotion and duplication: and those without resources in the first place – perhaps money, ability or knowledge – begin with a great disadvantage.

Companies are themselves small economic systems, and they share many of the characteristics of bigger groupings. A large number of distinct but interconnected activities are conducted within their boundaries; activities that are continually exchanging materials and services with each other. The boundaries are mostly clear and stable, while remaining fairly open to the outside world. So we can consider the pluses and minuses of markets and hierarchies *within* a manufacturing firm. We can question whether the exchanges of resources that take place in a firm should be conducted through market-like mechanisms, or whether they should be planned and directed by a few centrally-placed administrators. We can also wonder whether a firm's boundaries are set at the right points: would some of the purchased services and materials be better brought under the firm's command structure, or would some of the internal activities be better pushed out into a competitive market? Perhaps the new technologies will help companies find a new balance between markets and command structures in the future.

It is these issues that this section is intended to address.

The advantages of markets

A market is simply a system in which people or groups come together to make exchanges that are of benefit to them individually. Bargains are struck in which one thing is exchanged for another in a particular ratio: normally, one of the things is money, and the ratio is a price conveniently expressed in monetary terms. This price is the crux of the exchange because it summarizes the equilibrium, so to speak, that is reached between a seller wanting to get the most for the service or good he supplies, and a buyer wanting to sacrifice the least possible quantity of resources in order to obtain it.

If additional sellers of the same thing enter the market then, all other things remaining unchanged, the additional competition among sellers tends to drive the price down. If additional buyers enter the market, the price would tend to rise. Should, in the latter case, the price rise very far, then two distinct effects are likely to arise. The first is that, at a higher price, additional sources of supply will become economic, and the available quantity of the good might increase to satisfy the expanded demand. A rising price sends a signal to potential suppliers to search for additional sources of supply. If these extra goods become available, they will tend to counteract an increasing rise in price. The second effect is that some buyers could be driven from the market, no longer finding that the benefit they obtain from the good outweighs the price. The buyers remaining are those that are prepared to pay most for the good – those who therefore gain the greatest benefit from it.

In practice, the operation of markets can be complex.They often take on a recognizable structure: perhaps with few suppliers and many buyers, perhaps with many suppliers and one buyer. There might be one supplier or one buyer who is dominant because of the share of the market he controls. Or there might be large numbers both of suppliers and of buyers. On the basis of the market's structure it is arguably possible to deduce the conduct of the participants (who might form cartels, or exploit monopoly conditions, perhaps). It is even conceivable that one could go on to predict the performance of the participants once their conduct is known.

But these are, for the moment, unnecessary complications. The essential point about a market is its ability to allocate things whose supply is limited, to the buyers willing to pay most for them – the buyers who think they will derive the greatest benefit. If two manufacturing activities need the same raw material, for example, the one that can use the material to make the more useful products, with the more efficient processes, will make the most money. The people that operate it will therefore be willing to pay the higher price for the raw material. The market's mechanism tends to allocate all, or at least more, of the material to them. In other words, the market displays a degree of allocative efficiency.

This much could be done equally as well by a planner directing the flow of the raw material to the two activities in proportion, say, to the value they add to the material. There are some very common conditions, however, under which the planner (but not the market) would become overwhelmed by the scale of his task. The first is when knowledge of the benefits and costs of a good are distributed among all the bodies that provide or consume it. The planner would have to gather this knowledge and keep it up-to-date in order to maintain his allocative efficiency. The second condition is when circumstances are rapidly changing – when new uses for a commodity are being found, when existing uses are rising or falling in popularity, when new sources of supply are being discovered, and new production processes introduced. In the real world the planner's information-gathering task alone would be enormous. The job of processing this information and issuing commands to all the parties involved would be even more daunting.

In a market, the action of co-ordinating many independently-acting bodies is performed, impersonally, by the price. It is the price that summarizes how much more or less difficult things are to procure compared with others, adjusting every now and then to reflect new pressures of demand or new possibilities for supply. This information is exactly that needed for satisfactory allocation, and it is a piece of information that is broadcast throughout the market. Goods are allocated efficiently whenever the independently-acting participants tailor their consumption to reflect the difference between the benefit they get and the price they pay: whenever, in other words, they act sensibly in their own interest.

If, for instance, a producer found that the price of a raw material increased to such a level that his costs rose above his revenues, he would presumably stop buying the material and attempt to find another – or switch production to a product that could command greater revenues. If another producer had a much better use for the material he would most likely continue to buy it, even at the higher price. Without any one participant knowing of the circumstances of any other, the market manages to channel the flow of the material to where it is most useful.

Problems with markets

There are of course instances when what a market is supposed to do is not enough for gracious and civilized living, and there are circumstances in which it fails to do even what it is supposed to do.

For a start, the equilibrium that is eventually reached between buyers and sellers – the price-determining process – might be a lop-sided one if either buyers or sellers exert a disproportionate amount of influence. This tends to work against the ability of the market to reach a broadly satisfactory allocation of resources. Occasionally the dynamic processes taking place in a market never lead to a position of equilibrium. And it may be that the process of reaching an acceptable equilibrium is so slow that for most of the time the market doesn't in fact get goods to those willing to pay the most for them. Competition among participants is therefore an essential element in driving prices to levels that fully reflect the prevailing opportunities for demand and supply. Competition, however, has its own inefficiencies. Advertising, for instance, is sometimes a good way of conveying information to potential buyers and sellers, but it consumes resources in its own right, and is often intended to be manipulative rather than informative. And competition frequently leads to duplication – duplicated factories, duplicated administration and so forth.

Markets work poorly when there are spillovers associated with the goods or services exchanged within them. Producers tend to under-produce whenever they cannot appropriate the value due to their activities – perhaps because they cannot exclude non-payers from obtaining their products. Equally, they tend to over-produce whenever they avoid paying the full cost of the resources they consume – when their activities cause a polluted ecology or a hazard to health, perhaps, and they are not penalized as a result. The problem is usually one of inadequate property rights, where the ownership of a resource, or the entitlement to use it in a particular fashion, cannot adequately be defined.

The conclusion that we might draw is that the world is divided into two parts: the part containing things that can be satisfactorily exchanged in markets, and the part containing things that can't.

Markets and companies

There are two practical issues that are of interest here. The first is the continuing suitability of the boundaries that separate a firm from the outside world, and the need to adjust them to reflect a better way of applying resources to the production of goods or services. It could be that, as circumstances and technologies change, it comes to make better sense to buy a service from outside rather than generate it internally. Perhaps the ease of transmitting information on their needs to potential suppliers encourages a company's managers to think that they need less direct control over the way a service is generated. Perhaps these managers are beginning to consider that the presence of competition would reduce the cost of the service, or that removing it from the company's organization would enable them to focus more effectively on just those processes in which they have a clear, competitive advantage. This seems in fact to be the direction of current change: far from becoming ever larger, companies are in many cases getting smaller, and this trend is correlated with the use of computers.[11]

The second issue is whether activities that will always be carried out within a company might be better administered in a way that mimics the operation of a market, while retaining the sense of strategic direction and ultimate sanction over quality that comes from direct control. Cellular factories and internal markets are fashionable ideas, and interestingly, both are associated with new technologies, particularly new computer applications. Exactly why this should be so is explored in more detail a little later on.

Before that, we need to look at the factors that determine whether activities are conducted in markets or in hierarchical organizations. It will be by looking at the influences on these factors that the impact of technology on the balance between markets and hierarchies can be assessed.

Markets and hierarchies

The factors that determine where exchanges take place – whether in markets or hierarchies – are usually discussed by referring to market failures, or instances in which markets have come to be replaced by hierarchies. As will become evident this is more a matter of convenience, and perhaps historical trend, than logical necessity. In searching for opportunities to apply the new technologies we shall, in fact, just as much be considering the reverse process.

It is tempting to assume that hierarchies replace markets solely because of *inseparabilities*. For example, one manufacturing process might have to occur so soon after another that the two could only be conducted under the same roof (and, by implication, within the same company). To operate each process in a separate company, and have one company sell its products to the other, would be impractical because the processes are effectively inseparable. The idea of inseparabilities, however, does not really capture the essence of the problem. It is apparent that many companies indulge in several activities that could quite easily be separated from one another. In any case, factors such as physical proximity do not, by themselves, imply the need for a single control structure.

A more general and enduring approach is to consider *transaction costs*[12] – all the expenses, tangible and intangible, associated with performing commercial activities. There are two main sources of such costs: the resources spent acquiring the information needed to conduct a transaction, and the resources spent in the act of conducting the transaction itself – in bargaining, contracting and monitoring. To buy product components, for instance, a firm has to search for potential suppliers, determine the capabilities and quality of their products, find out whether delivery will be fast and dependable, establish the suppliers' financial stability and so on. Once it is ready to make a purchase, the buyer has to prepare specifications and deal with commercial documentation, arrange finance for the transaction, monitor the supplier's performance and the suitability of his goods. The process is a costly one for both provider and consumer.

Many of the costs stem from the fact that, in a market, knowledge is highly distributed. In most markets there is no single, central point at which a buyer can find out all the information he needs about suppliers, or a supplier about buyers. Both parties must go to some trouble transmitting and interpreting this information before transactions take place. The ability to make the most effective decisions – to pick the supplier who delivers the most cost-effective product for example – relies both on this information being accurate and complete, and on its recipient being able to process it fully. If there is a large degree of uncertainty about the information available, and if the person acting on it has a limited capacity to take all the information into account, the market won't achieve the efficiency discussed earlier. A company might buy an over-priced component because it does not have information on better products, or because its purchasing officer hasn't the time to sift through all the publicity material that aspiring suppliers send him.

Another source of transaction costs lies in the possibility of opportunism. When there are a small number of players in the market, and when one party to an agreement has more, relevant information than the other, that party can obtain a special advantage that is costly to the other. For example, when long-standing contracts are renewed, a buyer probably has fewer options than when contemplating the original contract – and the current supplier can exploit a position in which he probably has a substantial advantage over other suppliers.

Sometimes these costs can be avoided by changing the structure of the exchange, in particular by replacing the pattern of market transactions with a hierarchical control structure. This might economize on information and it can avoid some of the paraphernalia associated with contractual agreements. If, for example, a company were to bring the manufacture of a component under its wing – a process that had formerly been undertaken by another company – it might be able to determine the characteristics of the product of that process more accurately and more reliably. Information on the costs and technical performance of the process could be gathered speedily, and in detail, by the reporting mechanisms commonly used inside companies. Changing the output of the process could be accomplished by simply issuing new instructions to those carrying it out, rather than by lengthy negotiation with an outside firm. And there is less temptation for people to act in opportunistic ways because their actions are more visible to those with whom they work – they are more vulnerable to speedy punishment if their actions undermine the performance of the company as a whole. They do not, in other words, have the protection sometimes provided by the arm's-length relationship of the market.

Equally, however, companies can incur excessive costs in their own control structures. Information flows, in particular, can become delayed and distorted as they thread through an increasingly extended organization. Not only do straightforward messages suffer distortion when relayed from one person to another on successive occasions, but they lose much of their impact when they become aggregated with others as they ascend a firm's hierarchy. Still further, many of the messages are wrong when they start out: the inability of some accounting systems to trace costs to the activities that cause them means that such systems produce information that is wholly inaccurate before it is ever transmitted. Wrong messages usually get weeded out of markets because they lead to firms supplying products that nobody wants, or charging prices that fail to cover costs.

Consider a firm that operates an FMS and a conventional jobbing shop – each of which can produce the same complement of products. If the reporting systems do not identify the pattern of costs *caused* by each of the two possible processes, the firm will have no satisfactory basis on which to choose the more economical way of producing its outputs. Its decision will, perhaps, be based on maximizing utilization figures. In a market, in which the separate processes are carried out in different firms, products and services are exchanged to the advantage of both provider and consumer. If a process incurs a cost then this will undoubtedly be reflected in the price of that process's product: and if there is a better process that produces the same product then it will become predominant.

Ultimately, then, the criterion that tends to determine the pattern of exchange – whether a good or bad service is obtained in a market or by an internal organization – is the relative transaction cost incurred in each of the two cases. The processes by which transactions are carried out are like any other in that their cost is a function of the technology that can be brought to bear. When this technology is substantially changing, the balance of relative costs needs to be re-assessed.

Making markets

The demand model of Section 2.1 indicated that a manufacturer's ability to innovate and respond to more specialized desires will become increasingly important. But there are two reasons why hierarchical command structures might interfere with a firm's ability to innovate. The first is that in vertically-integrated firms it is easy for most people to forget about the firm's customers. They have no contact with their firms' product markets because they are surrounded by internal departments. There is so much organizational structure that outputs are obscured by intermediate goods and services. Maybe a firm's attentions turn inwards to such a degree that it becomes wholly unresponsive to the pressure of the changing demands. The second reason is that there is a danger that, even if a satisfactory understanding of product demand is achieved, there will be too great an inertia to make changes in time.

For the sake of illustating how computer technologies might take effect, I am therefore going to suggest that, for the most part, manufacturing firms have reached the stage where it is worth examining whether they could make more use of market mechanisms rather than hierarchies. We know that such mechanisms are not appropriate to every kind of activity, but then we are not concerned with every kind of activity: only with ones that lie close to the margin between the two types of structure.

The first attraction of market mechanisms is their ability to induce people to do the appropriate things with little explicit instruction.[13] They motivate people, even those who disagree that they are the most effective means of running a system, and they exercise a co-ordinating function that doesn't call for massive planning and massive control. They preserve a measure of individuality, and they rely on self-interest instead of the omniscience and complete rationality of a planner.

A market is an environment in which the participants generally perceive themselves to have greater control over their circumstances than in a centrally-directed system whose instructions (even if they are somehow correct) often appear arbitrary. Markets and competition provide a more compelling environment in which to change and improve than a rigid hierarchy. Markets should be good at dealing with change because knowledge remains local, close to the point at which it is applied to decisions about buying new products and exploiting new resources and processes. Planning essentially assumes that when a certain stimulus is applied to people or groups they make a certain response: knowing this, a planner works out what to do to achieve the response he wants. In reality, behaviour is not so deterministic, and there are many things (some impossible to observe) that cause people to react in different ways at different times to similar conditions. It is simply better to provide incentives than instructions.

A second attraction is the possibility that these mechanisms will allocate resources in a more effective way than many planners in complicated businesses can manage. It is the belief that planners just cannot capture all the relevant information they need in order to allocate goods and services in the most efficient way. It is the belief that innovation and responsiveness are especially well-served in decentralized structures in which decision making is close to its tangible impact. We want organizations that are close to good intelligence about demand and supply (that is, markets) and which are small enough to react to it.

If we can demonstrate that with new technology there is no longer a loss in efficiency by adopting market mechanisms in certain instances (because computers reduce transaction costs), we can think about obtaining these improvements without penalty. If market mechanisms become cheap once we use a database to bring buyers and sellers together, and electronic invoicing and payment systems to make transactions fast and cheap, we stand a better chance of overcoming scepticism about the reality of more diffuse benefits like stronger motivation and a better responsiveness.

What forms might these market mechanisms then take? The first is simply that a company subcontracts certain activities that it had previously conducted internally. These are likely to be activities well away from the core of the business in which a company may be expected to have a distinct advantage of some sort. A manufacturing company might sensibly consider obtaining its computer services from the market, but probably not its design work. It may well consider subcontracting certain manufacturing processes, or buying components in a state in which they are more nearly finished, but not if the efficiency or efficacy of these internal processes are so good as to set the company apart from its competitors. One has also to bear in mind that while subcontracting ancillary services is an evident trend, so is the increasing closeness of manufacturers and their suppliers of components and materials. This closeness need not lead to complete vertical integration within the boundaries of a single firm, but it suggests that short-term contractual relations are insufficient to obtain a satisfactory delivery and quality performance.

It may seem other-worldly to suggest that anyone within a firm is likely to pursue a policy in which the firm's direct command over resources is diminished, in which its boundaries are brought closer to its core. But there is nothing permanent about these boundaries, and the idea that they are determined by transactions costs suggests of course that they ought to move when transactions costs change. If you were to take the view of writers such as Jensen and Meckling[14] – that a firm is above all a legal fiction with which individuals separately form contractual relationships – then there is perhaps little to keep it together other than the sort of economies that differences in transaction costs yield.

An alternative is to retain an internal organization to co-ordinate the separate activities, but to place it in competition with contractors who can provide a substitutable service. The outsiders may not look especially like the insiders. But it isn't necessary that the organizational forms or even the products of the two are the same – only that one can be satisfactorily substituted for the other. For instance, over quite a broad range of applications, it is perfectly possible that suppliers of standard computer programs could compete with an internal computer department that writes one-off programs.

Another possibility is to use some of the mechanisms found in markets to help operations that remain internal to the company. For example, different departments that can provide the same output might be called on to bid for work in competition with one another, and made to face the consequences either of doing no business or of doing business that would essentially bankrupt them. This principle might be applied to manufacturing cells, for instance. One can conceive of two assembly cell managers, perhaps, indulging in a continual race to outdo each other in terms of controlling their cell's costs, steadily increasing their quality and reducing their lead-times.

Some preconditions

There are a number of elements that would be needed to make this a practicable approach. First, the costs that competing providers incur should, on the whole, be real ones. In other words the charges they pay for the resources they use ought to be opportunity costs: they should reflect the sacrifice made by denying these resources to other, potentially profitable activities. The idea of setting capital costs at a fixed proportion of a notionally direct labour cost would be wholly inappropriate. Any consumption of resources within the control of the provider must be unequivocally set against his finances. Of course he will benefit (or perhaps suffer) from such services as administration, accounting and so on which *might* not be traceable to specific consumers. Such overheads are somewhat akin to a general tax. But the central idea is to charge consumers of resources, whenever it is possible, for exactly those resources they consume.

A second element is that providers must be put in a position from which they can exercise discretion and responsibility. They must be allowed a good measure of freedom to decide how to produce their products, and they should not be forced into transactions in which they don't want to participate. Equally, they have to be held accountable for the state of their finances, and the providers of these (the company as a whole) must exact a charge that it is appropriate to the terms of the funding and the risk associated with it.

It is probably a good idea to look out for the common forms of market failure: for providers exploiting monopoly positions and spillovers, for instance. But the quasi-market need not be perfect, only better than rigidly hierarchical planning. And those phenomena such as monopoly that are generally felt to be bad for the system may, on closer inspection, be quite acceptable. There is a case for saying that it is the relative power of consumer and provider that is important, not the number of suppliers, or the number of customers. In any case, the prospect of having some unique advantage, of being able to extract monopolistic benefits, is what often appears to drive suppliers to take considerable risks in development. Without a monopoly, a supplier might feel no great incentive to innovate.

As an aside, it is interesting that the commodities that yield scope economies are a type of public good, because the conditions that make them so (non-rivalry in consumption and non-excludability) suggest that markets in such commodities will not work properly. This means that we might not be able to use such mechanisms for many information-producing activities.

Technology and progress

What is the role of technology in making market systems feasible? There are two main components of transaction costs (the expenses of information and of contracting) and it is convenient to use these as a starting point for identifying promising technological applications.

First, information costs can be reduced by making information more widely available and more easily retrievable. An obvious mechanism to use is a global database: global in the sense that its contents are accessible to all who can make legitimate use of them. Associated with this database there need to be facilities that minimize the costs of locating information – facilities that allow people to navigate their way through the contents, and facilities that allow people to specify the information they would like to retrieve.

Databases plus data networks can provide cheap broking services by bringing together buyers and sellers. If there is a central system that records the capabilities of producers and the desires of consumers, there is a way of matching one with the other without either having to conduct a lengthy search process. This also makes it easy for participants to compare the relative merits of what other participants are offering: it helps make sure that the exchanges that do take place are the most suitable ones.

Databases also make it easier to reduce uncertainties in the technical properties of a service or a product. For example, the ability of two manufacturing cells to tender for business rests on the ease with which they can get access to product descriptions (such as specifications and drawings), and perhaps process instructions. Equally, the people that make the decision about allocating business to the cells will need to have information on the cells' past performance: whether they delivered on time, to specification, within the tendered price.

Reducing the second component of transaction costs – that of effecting the transaction itself – is plainly the rationale for electronic data interchange. Ordering, invoicing and transferring payments electronically promises market mechanisms that are cheaper, quicker and more certain than they were in the past. Companies may even contemplate removing the invoicing stage completely, automatically paying for goods as they are delivered. This reduces the funding that delayed payment provides, but eliminates the cost of processing invoices. The rather intangible nature of the motivation benefit that one might look for in market-like systems makes such schemes all too easy to criticize when administrative costs appear high. So the need for quick and cheap transaction systems is especially important for internal quasi-markets, for here the value of the transactions themselves is likely to be that much smaller.

These examples of the role of computers in commercial transactions are summarized in Figure 2.4.

An example of a scheme in which transactions *within* a factory are automated is described in a paper by Shaw.[15] In this scheme, a number of flexible manufacturing cells can bid for business against one another, and the work is

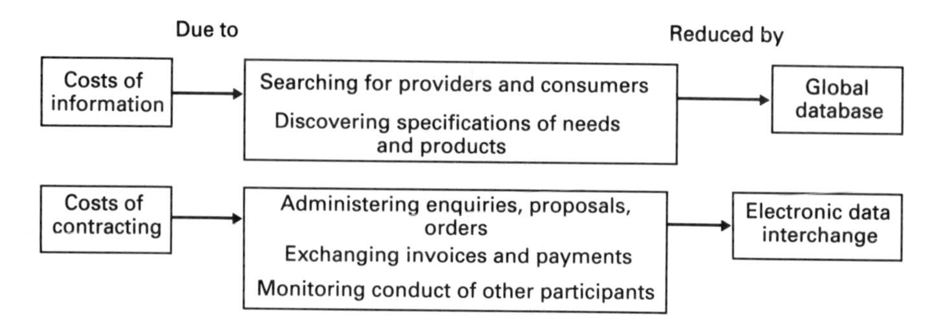

Figure 2.4 Computers and transaction costs

allocated on the basis of the bids they return to the system issuing the enquiry. The enquirer might, for instance, give the job to the cell that promises the earliest delivery date, although, in a scheme of this type, it is perfectly possible to use another way of choosing among bidders. The cells maintain a record of work for which they have bid, so that this can be taken into account whenever they seek new business. They also plainly have knowledge of the resources they must call upon in order to perform a job, and the price (or time) needed to acquire them; this is incorporated in their bids.

Shaw found in a simulation that the performance of this distributed scheme was significantly better than that of a system which scheduled work by calculation at a central point. In addition, he suggested that the distribution of planning and decision making leads to better reliability, since the scheme as a whole does not fail suddenly when any one of the participants breaks down. It is more extensible because the control structure remains unchanged as new cells are added to the scheme, and it is also more cost-effective since much less powerful information-processing systems are needed. Rather surprisingly, the distribution of acitivities means that there is less communication activity than in a centralized scheme: this is because it is only bidding and allocation messages that need to be exchanged. In a central scheme, by contrast, the cells have to transmit a good deal of information to the central scheduler about the products they can handle, the resources they need to do so, the time it takes them and so forth. (Think of the information about every work-centre in a factory that has to be recorded and maintained in an up-to-date condition by a centralized MRP system.)

This scheme does not fully illustrate the operation of a market in the normal sense because work is contracted to participating cells on the basis of a very limited index of performance, such as the earliest achievable due date. But the fact that it co-ordinates a factory's work better than conventional scheduling procedures suggests that much of the value associated with client–contractor structures, like distributed decision making and local information storage, can be gained without doing anything too elaborate. It is an encouraging and practical model.

Even if the mechanisms associated with markets are not adopted, the ability with computers to replicate human information-interpreting expertise suggests that a firm can substantially flatten its hierarchy and push down the decision making process to lower levels. The information filtering and adulterating activities of intermediate levels can certainly be replaced by technology, a fact that has been frequently noted – if not acted upon. Removing parts of a hierarchy goes a little of the way towards promoting motivation efficiencies, but it is worth bearing in mind that the difference between planning and markets is one of kind, and not simply one of degree. The attraction of markets is that their participants receive no instruction whatever: their actions, in the right setting, have enough characteristics of their own to bring about a high level of co-ordination. It is the replacement of instruction by incentive, as much as that of centralization by decentralization, that makes the mechanism an attractive one.

However, it would be dogmatic to suggest that this is inevitably the right direction in which change should take place. In specific circumstances it may in fact prove to be better to centralize and organize activities hierarchically. With computer systems that can avoid the distortion and attenuation of facts and figures as they pass through the firm, some of the hierarchical transaction costs are reduced. The essential point is that new technology (especially information technology) warrants a re-examination of the boundaries between the two forms of organization. The concept of transaction costs will be a good guide to this re-examination, and it helps reveal the role of technology in improving the conduct of exchanges.

2.4 Summary and suggestions

The purpose of this chapter has been to discuss some of the themes that underlie the use of specific technologies in manufacturing. This ought to provide a way both of identifying promising developments, and of testing the rationale suggested for specific projects against consistent patterns of progress.

The first section looked at the changing nature of customers' demands in the markets for manufactured goods. This suggested that a firm's focus should evolve through a series of phases – phases in which the emphasis is placed successively on:

- minimizing costs;
- maintaining quality;
- producing products in variety; and
- exhibiting the will to innovate.

In many cases the most current interests are in variety and innovation. Whatever the nature of these new needs, however, it seems obvious that they should be reflected in the developments a firm pursues within its factories: new process technology has a considerable influence on the ability to meet new product demands.

The second section looked at how flexible systems can be exploited in order to achieve product variety in an economical fashion. This hinges on finding economies of scope – on being able to share inputs to the manufacturing process between the production of different products. The test of a sharable resource is that it should display two characteristics:

- its consumption in one activity shoudn't deny consumption to others (non-rivalry); and
- the consumption of the resource should be possible at very little cost to its consumer (non-excludability).

It is hard, in practice, to measure properties such as variety and flexibility in a credible manner, but the idea of scope economies provides a precise enough model for reasoning about these issues in a qualitative way. It helps translate flexibility into commercial characteristics, and it suggests a number of technologies that ought to be worth a more detailed inspection. Among the technologies that promise such economies, for instance, are flexible manufacturing systems and knowledge-based systems.

The third section looked at how a firm might reconsider the boundary between markets and hierarchies. The main incentive was to find organizational structures that would be better at innovation, and in some respects market mechanisms look more suitable than command hierarchies. Since new technologies can be used to reduce transaction costs in market-like systems, this is now a good time to re-assess which mechanisms are the most suitable for the allocation of resources within a business. In particular, computers can be used to reduce

- the costs of information (perhaps databases to help search for buyers and sellers), and
- the costs of contracting (perhaps EDI for automating orders, invoices and payments).

There are at least two ways of applying these principles. The first is in a passive fashion, testing proposals for specific technologies that have already been identified against the rationale behind either scope economies or transaction costs. For example, if a technology promises flexibility, you might ask what this flexibility means in terms of the variety of products that it will help the firm produce. Is it the type of flexibility that would let the firm undertake processes that it cannot do at present, or does it simply offer a number of ways of carrying out the same processes? Does the firm have the will to make use of this flexibility, and are conditions in the firms's markets such that a greater product variety can successfully be exploited?

If a new technology is said to reduce transaction costs, a firm should be asking whether its organization ought to change to take advantage of such economies. Global databases are of little value to an organization that continues to restrict access to information on a traditional basis: if an engineering department continues to keep designs private until it issues them by paper-based systems, perhaps. It would, similarly, be questionable to refine the methods of a hierarchy if greater gains are to be had by adopting markets. Decision support systems based on aggregating information over successive levels of a hierarchy would not be suited to monitoring the transactions among cells in a quasi-market.

The second approach is to apply these basic ideas in a more active manner. A firm might use the principle of scope economies to form a strategy for internal change. This would be especially appropriate if it accords with a marketing policy intended to increase the variety of products the firm will market. Customer's demands determine the extent of the variety in a firm's output, and this needs to be matched by the flexibility of its processes. The degree of this flexibility is the focus of the strategy. Targets for flexibility may not be quantified, but they need to be clear enough that people can at least describe them in terms of their operational effects.

Strategies for markets and hierarchies are more difficult to find because there are a good many issues that influence what is a very fundamental choice for a firm to make. It cannot sensibly contemplate putting even a small, internal service out into the marketplace until it has some idea of the market's effectiveness – whether it avoids some of the more serious imperfections, for example. Such changes are also likely to be difficult to square with the petty politics of an established organization.

Notes and references

1 Bolwijn, P. T. and Kumpe, T. Manufacturing in the 1990s – productivity, flexibility and innovation. *Long Range Planning*, **23**(4), 44–57 (1990)

2 Gold, B. Changing perspectives on size, scale, and returns: an interpretive survey. *Journal of Economic Literature*, **19**, 5–33 (1981)

3 Talaysum, A. T., Hassan, M. Z. and Goldhar, J. D. Scale *vs* scope considerations in the CIM/FMS factory. In Kusiak, A. (ed.), *Flexible Manufacturing Systems: Methods and Studies*, North Holland (1986)

4 Panzar, J. C. and Willig, R. D. Economies of scope. *American Economic Review*, **71**(2), 268–72 (1981)

5 *Op. cit.*

6 *The Economist*, Information Technology Survey. A game everyone can play, 16th June 1990, 30–1

7 Slack, N. Manufacturing systems flexibility – an assessment procedure. *Computer Integrated Manufacturing Systems*, **1**(1), 25–31 (1988)

8 Kumar, V. On measurement of flexibility in flexible manufacturing systems: an information-theoretic approach. In Stecke, K. E. and Suri, R. (eds.), *Proceedings of the 2nd ORSA/TIMS Conference on Flexible Manufacturing Systems*, Elsevier, Amsterdam pp. 131–43 (1986)

9 Kulatilaka, N. Valuing the flexibility of flexible manufacturing systems. *IEEE Transactions on Engineering Management*, **35**(4), 250–7 (1988)

10 *Op. cit.*

11 *The Economist*, The incredible shrinking company, 15th December 1990, 89–90

12 An extensive discussion of the subject is provided by Williamson, O. E. *Markets and Hierarchies: Analysis and Antitrust Implications*. Macmillan, New York (1975)

13 Hayek, F. A. The use of knowledge in society. *American Economic Review*, **35**(4), 519–30 (1945)

14 Jensen, M. C. and Meckling, W. H. Theory of the firm: managerial behaviour, agency costs and ownership structure. *Journal of Financial Economics*, **3**, 305–60 (1976)

15 Shaw, M. J. A distributed scheduling method for computer integrated manufacturing: the use of local area networks in cellular systems. *International Journal of Production Research*, **25**(9), 1285–303 (1987)

3 A role for computer technologies

My father worked for the same firm for twelve years. They fired him. They replaced him with a tiny gadget this big. It does everything that my father does, only it does it much better. The depressing thing is my mother ran out and bought one.

Woody Allen *The Nightclub Years, 1964–68*

The purpose of this chapter is to discuss the forms that computer-based technologies take, and the main economic issues that follow from them. The arguments are informal, and remain for the time being very general: they are about broad classes of system rather than specific instances.

The first section looks at some distinctive characteristics of the technology, and the way these influence its application in manufacturing firms. The second discusses the significance of what you might call the technology's raw material – information. It suggests the things we know (or want to know) about it, what forms it can take, and how it can best be managed. The third section considers how this reasoning lies behind a classification of the benefits we might expect technologies and applications to exhibit. This classification is intended as a starting point to the process of searching for technological opportunities. It is not, however, sensible to treat such a classification as being definitive, because it is plainly important that firms introducing new systems exercise a degree of imagination in doing so. They are, for the most part, the experts in the particular circumstances in which their operations are conducted. It is equally important that the manner in which the technology is applied reflects the changing pressures on an organization, and its changing views of where its priorities lie.

3.1 Primary characteristics

It is difficult to draw a clearly defined boundary around the scope of what you would call advanced, computer-based manufacturing systems. I shan't worry too much about definitions since they are often contentious and liable to become quickly outdated. The concern here is with manufacturing systems in the broadest sense – with systems intended to support any of the operations in a manufacturing firm; its engineering and commercial processes as much as its production activities.

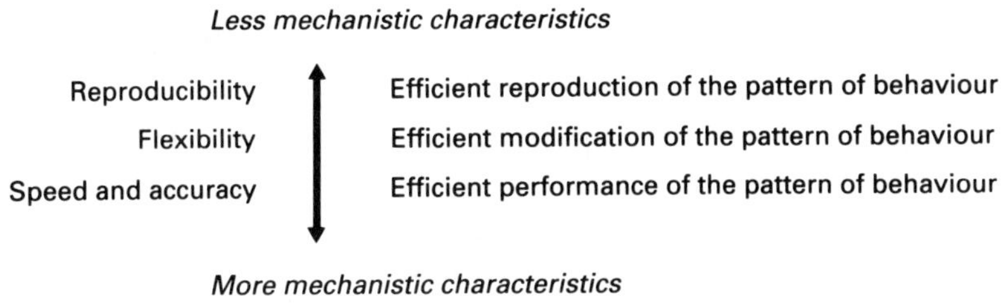

Figure 3.1 The distinctiveness of information technology

The focus rests on systems that embody information technology. Important as they are, technologies concerned with forming materials, reacting materials, sensing, actuation and so forth are ruled out of the scope of this discussion. It is what computers do for manufacturing systems that is of interest here – their ability, essentially, to capture know-how in the form of software and to apply it very quickly and very precisely. But this ability is something more than performing a particular task with great speed and accuracy, since computers are not (unlike, say, Swiss watches) highly intricate mechanisms alone. The fact that systems which use computers put the knowledge needed to accomplish their tasks in a malleable form suggests that they offer an entirely new way of doing things.

This malleability doesn't just mean that computer controls can be updated at a fraction of the expense of modifying mechanical controls. It means that know-how can be reproduced countless numbers of times, cheaply and at short notice. An algorithm or a set of rules can be used as effectively by the one-hundredth machine that stores it as by the first. In this respect computers vastly exceed the capacities of people, who can only acquire new knowledge with considerable concentration and after a good deal of time.[1] So as well as speed and accuracy, computers confer flexibility; and as well as flexibility they offer what you might call reproducibility (Figure 3.1).

It would of course be a mistake to think that computer-based systems are perfectly flexible, because there is always a finite cost to reprogramming. This is especially so when attempting to modify complex, ill-structured systems which become increasingly difficult to fashion to changing demands. Nonetheless it is usually the case that the modification of software is much less costly, for an application of a given complexity, than that of purely mechanical or electrical equipment.

All this comes at a cost of course, and a part of the cost is that firms wanting to make best use of the technology have first to understand themselves. They have to understand the complexities, compromises and inefficiencies of their own organization if they want a malleable tool to fit their particular mould. And they have to face up to the consequences that taking a good look at themselves might have. Often the result is an organizational upheaval of some sort whose effects are as profound as those of the technology itself.

This process of re-orientation carries perhaps the biggest price associated with introducing computer-based technology – a price even greater than that of buying computers, application programs and training courses. For the most part, the rest of this chapter assumes that this price will be paid: that the introduction of a technology will be conducted in a sensible way, with few corners cut. When the talk is of benefits and value it is really of potential benefits, and the value contingent on doing things thoroughly. This shorthand may sometimes make the discussion seem optimistic – but it avoids having to make repeated qualifications to the main thread of the discussion. Many of the practical issues that must be addressed if systems are to yield the value attributed to them are discussed in later chapters.

Some appealing properties

The first characteristic of the technology, then, is that it has computing at its core – not simply to perform tasks peripheral to the purpose of a system, but to carry out its essential functions. Moreover, the interesting part of this computing is the software. This software usually has both a specific pattern of behaviour that follows from the particular programs it runs, and a general way of solving problems that depends, in addition, on the techniques and mechanisms that underlie these programs.

Consider, for example, an expert system that helps a product designer to make sure his creations meet a few basic design-for-manufacture criteria. At any particular moment it has a program – the set of rules and knowledge of other sorts that people have told it to record from time to time. The way in which this knowledge is recorded and applied, however, depends on the programming languages, the shell programs and other elements on top of which the knowledge about design-for-manufacture is effectively overlaid. For the sake of brevity, call the knowledge specific to a particular job the *application* and all that underlies it the *technique*. Then both the application and technique determine the system's behaviour and its value and, in particular, the second sets some limits to the capabilities of the first. Our concern here is almost exclusively with the application and its outward effects rather than its internal form. But it is sometimes practicable to make generalizations based on the technique rather than the application, and we can then apply this line of reasoning to a number of different applications. We might argue that a benefit of *any* database is that it potentially offers access to a consistent body of easily-updated information, over wide geographical areas, to large numbers of people.

This separation of application and technique has a number of economies. It allows software writers to specialize to some extent, thereby reducing the need for them to duplicate work. And it allows the general-purpose element (the technique) to be the subject of standard definitions. The use of SQL as a common data manipulation language, for instance, means that applications are more readily copied to different types of computer. It means that the market for application programmers is a larger and more informed one, and it suggests that in any particular firm there is more scope to re-use old modules in new programs.

The fact that new systems embody the means of manipulating information suggests a second characteristic: the capability of abstraction. In other words, programs can be written to discover useful and interesting properties; they can select and process pieces of information that will yield a particular view of the activities under their control. For instance, it is generally possible to obtain sophisticated diagnostic information from machine controllers, helping people to find faults quickly and accurately. The information needed for statistical process control can often be gathered readily from process controllers, and decision support modules can be added to accounting systems. It is unfortunate that the use of computers is commonly associated with too much information – with vast and detailed listings that cannot possibly be digested by the people for whom they are intended. A computer is in fact the ideal mechanism for spotting patterns, identifying associations, and extracting terse and relevant meanings from large volumes of elementary data. Their ability to help decision makers abstract from this data is, potentially, one of their most attractive characteristics.

A further characteristic of computer-based technologies is that they embody a considerable potential for further innovation. They can be applied, often without too much difficulty, to unexpected tasks. Once people latch on to the idea that something is programmable, there is usually a feverish desire to experiment, and to tailor a system to the personalities that put it to use. This potential means, however, that it is difficult to predict the effects of a system from the nature of the technology alone.[2] The ideal approach is perhaps to plan in detail the manner in which, in the first instance, a new system will be applied to support the firm's operations – but at the same time to leave open the possibility of further adaptation to changing needs and desires. This means, unfortunately, that much of the value of the technology will be very hard to predict in advance. It reflects the presence of growth options, an idea discussed in Chapter 6.

Less appealing properties

One characteristic of advanced technology is of course that it remains novel to many firms. The depth of experience that enables organizations to make decisions automatically and repetitively hasn't yet been acquired on a widespread basis. Typical among such decisions is that about whether, and how, to invest money in buying the technology. And, preceding this, how to assess its worth. In fact, to think that this experience will come with time is perhaps being too hopeful, because another characteristic is the rapidity with which the technology is changing. Whatever the approach a company takes to making decisions about this technology, it will need to be one that copes well with change if it is not to become rapidly obsolete. And if there is never the time to develop rules of thumb in the way that was possible when technologies were changing more slowly, the new way of making decisions is likely to remain close to basic principles.

A further problem commonly associated with computer-based systems is the difficulty of escaping from company-wide effects when they are being introduced. Much as some firms try, it is mostly very hard to keep the influence of new systems within the confines of a single, well-defined department. Think of the places from which information is needed for computer-aided testing, for example: schematic capture systems, PCB layout programs, process planning systems, a test specification database, and the MRP system, perhaps. The key element is obviously information. Computer-based manufacturing systems need dispersed pieces of information just as much as highly localized, physical entities like raw material,

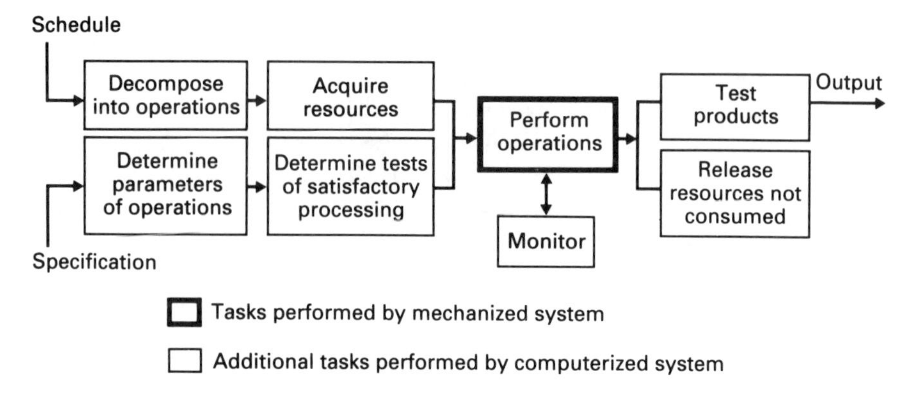

Figure 3.2 The reach of information processing systems

an electricity supply, and a person to tighten a vice and press a button. Their demands for resources extend much further away from the point at which a simpler machine might have done its work. They cannot work without schedules, specifications and so forth. Figure 3.2 is rather a simplistic sketch conveying the much greater reach of systems that incorporate information technology.

A final difficulty with assessing many computer-based systems is that their effects are several stages removed from the quantities ordinarily used to measure a firm's profitability. To understand how something like flexibility takes effect on financial performance, you would have to go through several stages of reasoning. Maybe flexibility is the ability to carry out quick product changeovers: this means smaller batches and therefore more concurrent operations: this in turn ought to yield shorter lead-times, which finally leads to bigger revenues because customers favour quick deliveries. This rather tenuous thread of causes and effects contrasts with the nature of simple cost-cutting exercises, whose impact is obvious and, for the most part, easily forecast. And not only is the chain lengthier, as it were, but it is also broader. Smaller batches perhaps mean better product quality as well as shorter lead-times; but they also need better mechanisms for synchronizing material flows. The benefits are therefore contingent on other developments taking place in the factory. The way in which this affects investment appraisal will be discussed in Chapter 5.

Trading certainties

Some writers have suggested that the main motivation for introducing computer-based systems is to reduce the enduring uncertainty in a firm's operations, even if this is done at the expense of increasing short-run uncertainties during the period of a system's introduction.[3]

Uncertainty has a number of sources. Processes are sometimes of a rather variable quality and machines suffer breakdowns at random times. People behave in unpredictable ways. And a company can never have perfect information about all the processes taking place outside its boundaries: prevailing levels of economic confidence influence levels of demand for a company's products, and they influence prices in the markets for raw materials, energy and purchased services.

The costs of uncertainty become evident in various forms. There are the costs of disrupted or faulty production, for instance: scrap, rework, missed delivery dates (and a loss of goodwill or the invocation of contract penalties). There is the possibility of shipping faulty products and incurring warranty costs. Also, there is the cost of any insurance the company attempts to take out against such problems – large inventories to make up for scrapped products, large cash balances to pay for additional materials or services, reductions in product variety and less frequent changes of product. The last two effectively mean that a company forgoes some of the revenues it would otherwise command: this opportunity cost is the insurance premium, if you like, that the firm pays to reduce the likelihood and impact of disruption.

New technologies can help in several ways. First, they might be used to deliver greater volumes of more accurate information to production processes – better specifications, more realistic schedules, more thoroughly verified part programs and so on. Second, they make these processes inherently more predictable and accurate. This is most apparent on the shop floor where electronic controllers for machines tools and process plant can frequently achieve much better precision and repeatability than human controllers. It is also true of computer-aided engineering tools: simulation improves the design of electronic systems, and drafting tools sometimes improve a designer's ability to visualize the form of his products. The new technologies help implement statistical production monitoring, and they provide the flexibility needed to tackle variety at a reasonable cost. Because this variety can be better managed, and manufacturing systems tailored to the production of particular products in a relatively short time, there is less need to reduce variety and innovation in order to avoid uncertainty in the manufacturing process.

Some of the new technologies are also intended to reduce working capital (particularly stocks of work-in-process), although the ability to do so probably depends at least as much on adjusting the nature of the manufacturing organization as on technology. In effect, one is buying fixed capital in place of working capital – adopting increasingly roundabout (that is, capital intensive) methods of production. This has been a notable characteristic of industrial progress in the past.

Effectiveness and efficiency

The efficiency of a process is of course the ratio of its useful outputs to its inputs. The fact that the outputs are useful means that efficiency must embody some measure of the process's ability to deliver a desired product, as well as an ability to deliver it without consuming a large quantity of resources. However, it has become customary to look upon efficiency as conveying only the ratio of measurable (and not necessarily useful) outputs to inputs, and as a result we have to introduce the idea of effectiveness. Effectiveness, in other words, is concerned with doing the right things, while efficiency is concerned with doing them well.

The assumption will be made in the subsequent discussion that the best judge of whether a firm is doing the right thing is the market for its products. It is reasonable to suppose that people buy a firm's products only to the extent that they reflect their preferences, and only to the extent that they are not too highly priced. Hence the test of whether the company is doing the right thing is the health of its outputs (in quantitative terms), or of its revenues (in financial terms). So the question as to how new technology improves effectiveness can be rephrased as a question which asks what effect the technology has on revenues. This new question is rather easier to answer.

The test of whether a company is conducting its operations well is usually the number of inputs it consumes in order to produce a given output. The inputs are sometimes distinguished from one another in order to examine the efficiency with respect to a particular input. But this separation is often suspect because there can be a degree of substitutability between inputs: this means that as the efficiency with respect to one increases, that with respect to another may well decrease. Buying more machinery to make do with fewer people does not, necessarily, represent a greater total efficiency. However, provided that we are careful to avoid this, questions about the impact on efficiency can be rephrased as questions about the impact on a firm's cost base. These, again, are a little easier to answer.

So whenever in subsequent sections there is a need to classify the benefits of a technology, a good starting point is to examine how it affects costs and how it affects revenues. This has still to be done with some recognition of the limitations: costs have an impact on prices (the exact nature of the relationship is a matter of some controversy), and efficiency therefore influences effectiveness. If, moreover, you believe in economies of scale (again a matter of controversy), effectiveness influences efficiency. But, acknowledging this complication, costs and revenues represent a rudimentary way of understanding a development's financial impact.

What sort of effects might occur on the revenue side? A firm's output plainly depends on the nature of the products it delivers, so any property of these products that is of material interest to customers is likely to have an effect on revenues. Qualities such as the product's function and design are important: so, normally, are its fitness for its intended purpose and the speed and dependability of its delivery. The emphasis that manufacturers commonly place on quality and lead-times is presumably a reflection of the extent to which these properties influence the amount of goods a firm can sell. Similarly, the increasing importance attached to the time it takes to bring new designs to market demonstrates the need for a capacity to innovate. The way in which customers trade off one property against another, and whether they do so at all, lies in the province of marketing theory. But all firms have some knowledge of which attributes are the most important in their own markets, and in those markets they have the mechanisms to test their hunches about their customers' preferences.

To some extent, revenues also depend on the structure of markets: the numbers and sizes of buyers and sellers, how easy it is for firms to enter and leave the market and so on. Traditionally, the view has been that the more concentrated a market is – the fewer suppliers there are, for instance – the more profitable it is for these suppliers. Whether structure really influences performance, and whether, if it does, this is a socially damaging phenomenon is the subject of continuing debate.[4] But it makes sense that the investigation of a technology's revenue effects should include some consideration of market structure. You might ask whether that technology could increase barriers to other firms wanting to enter your market, make it more costly for customers to switch to another supplier, or alter the balance of power in the company–customer relationship.[5] Providing customers with a computerized database about a firm's products, or better still a knowledge base, might make it much harder for companies without a history of expertise in an industry to gain a share of the market. It might also make it harder for customers to give their business to another supplier already in the market.

The impact on the cost base is usually more clear-cut, especially when systems are introduced specifically to save costs. A firm might attempt to reduce its consumption of raw materials by using nesting programs, or its energy consumption by using environmental control systems. But new systems obviously increase other types of cost, since one has to acquire machines, software, training, consulting services and so forth. If these things are bought from outside it is usually easy to quantify their costs because they have an explicit price. In other cases their costs are less obvious. Frequently, for example, a company will use its own employees to carry out technological developments and, because this does not involve transactions in which cash is exchanged for specific tasks, the opportunity costs are hard to measure. This type of work also involves elements, like risk, that ultimately carry such costs as the additional charges that the firm's financiers will make for the funds they supply. At the point at which these costs are incurred they are not traced back to specific developments, but reflect the nature of the firm's operations in aggregate. It is therefore hard to know just how big a risk premium should be added to the costs of a specific project.

These themes are elaborated upon in subsequent chapters.

3.2 Information and processes

Having said that the concern here is with computers, it is now worth taking a look at the stuff to which computers are applied – information. It is the fact that systems such as CAD-CAM, FMSs and so on deal primarily with information that makes their reach so wide and their introduction so complex. No longer does a firm want to put extended chains of technicians and administrators between drawing board and machine tool. It wants to connect directly whatever holds the design to whatever fashions the product. Bypassing people-intensive activities in this way is bound to be a tricky issue, and the evaluation of its effects will warrant a good deal of analysis.

Types of information

One of the more fundamental divisions is that between what you might call *prior* knowledge and *local* knowledge. Prior knowledge is information that is received by study or experience, and it contains general ideas that can be applied in situations which might not have been envisaged when the information was acquired. It is the type of knowledge that most people manage to absorb on formal training courses and during their general education. For instance, it is prior knowledge to know that the equations describing how certain types of robot must be actuated to achieve certain movements have singularities under specific conditions. An engineer who knows this will realize what to do if, when teaching a robot, it comes to a halt in mid-movement.

Local knowledge is that of specific events and situations. It would be local knowledge to know that a particular robot is being used to carry out a specific process – and perhaps that it breaks down whenever its workpiece is presented to it in a particular orientation. Traditionally, of course, people with much prior knowledge have been accorded greater status and material reward than people with much local knowledge, all other things being equal. This doubtless reflects the fact that gaining prior knowledge requires more investment on the individual's part, since he is often removed from the workplace and the source of a livelihood while acquiring it.

Of course both local and prior knowledge are necessary to accomplish anything of substance. But the main advantage of prior knowledge is that it frequently economizes on information. Knowing a law of some sort to be applicable to a large number of situations often saves knowing many fragmented details about those situations. Broadly speaking, it is better to understand and encode prior information in our information processing systems if this can serve as well as local knowledge in any one situation. It might even be better when it serves slightly less well in certain situations if, overall, the savings outweigh the costs. Having prior knowledge also, of course, gives people a degree of control over their future: they can make generalizations and form expectations about the future that for the most part will turn out to be reasonably accurate. This is clearly essential to all activities in a manufacturing firm. Provided that it is relevant and assigned due priority, scholarship is undoubtedly a good thing.

A similar distinction can be made between information composed of rules and that composed of facts. Again, recording the first is more economical than the second, in the sense that one rule can impart information about many situations, whereas one fact does not.

Although the distinction is not as clear-cut when explored in more depth, it is commonly the starting point for the division between databases and knowledge bases. Databases are collections of facts with a very narrow applicability. These facts have to be interpreted in some way by another mechanism – usually a person – in order to determine an appropriate course of action. Knowledge bases usually incorporate some capacity to reason about facts, and to derive sensible courses of action from them. This reasoning can be expressed in the form of procedures which spell out how the facts are combined and manipulated to yield a suitable result. It can be expressed in the form of rules and associations, or in the form of templates of typical sequences of events or circumstances. Whatever the form of reasoning, the cost of obtaining rules rather than facts is, like prior knowledge, likely to be high. The economic sense of rules lies in their wide applicability, and once encoded in a form manipulable by a computer they can be replicated and disseminated virtually at will.

The issues quickly become more complex, however. There are, for example, rules about rules in many applications. Such a rule might say that the requirements of design-for-testability should precede those of design-for-assembly whenever they indicate contradictory product qualities. Using this type of information is again an economy in that it saves having to record each combination of rules in which one has priority over another. There are also stereotypes: a form of information which suggests how something will appear or behave, while leaving open the possibility that in practice it will occasionally behave differently. This economizes on the mechanisms used to manipulate information because rules can be applied more often than not. The only specific information that is needed is a small set of exceptions. Suppose, for instance, that in a particular production process it is normal to have a heat treatment operation at some point, but that when one, special material is used this proves not to be necessary. The information that records this need only take the form of two rules – a generality, and one exception – instead of separate rules for every possible material.

All computer-based technologies need nearly all types of information, although they may only *explicitly* record the fact that they use one or two of them. On the face of it, straightforward databases simply store local knowledge that is entirely specific to particular circumstances. But plainly these databases embody information more general than the facts themselves: some rules about the ranges that such facts can lie within, some of the interrelationships (such as the rule that they must all be distinguishable), and the structures that specify how they are laid out on the page or screen to the best effect.

However, the relative proportions of these different types of information change with the purpose of the system. Herbert Simon[6] describes the history of information processing as beginning when our ancestors found that their poor short-term memory slowed down the process of doing sums. At the same time, the inaccuracy of their long-term memory made them forget important bits of know-how: so, in the mechanism of writing, they found a way of storing symbols outside the brain. More recently, nearer ancestors found that they could copy information cheaply and indefinitely by printing. Still more recently, they found that information could be transmitted over large distances. Finally, with automated systems, it was realized that information could be processed in an active sense – by recognizing patterns, making decisions and so forth. It is for the last of these applications that prior knowledge and rules are the most useful raw material. Expert systems fall mainly in this last category. Databases on the other hand perform only the more rudimentary functions of storage, copying and transmission.

What do we (want to) know about information?

It is not enough simply to characterize information as being of a particular type. To make sensible decisions about how, and whether, it can be put to use a good deal more needs to be known. For a start, information has a cost like any other commodity since the means used to acquire it and transform it into a useful form are in limited supply. In situations where information is marketable this is made obvious. Text books conveying prior knowledge, car price guides giving us aggregated facts, access to databases about people who are thought to be credit risks and so forth all have a finite price. There is no reason to think that information that lacks an explicit price is any less expensive. Since in fact it is often only useful to a handful of people within a single company, its costs are attributable to far fewer decisions – and the cost for a given decision is, as a result, likely to be much higher than that of traded information.

An important component of this expense is the time of the person who attends to the information, hopefully as a preliminary to making a decision. This carries a cost equal to the value that is sacrificed by taking the person away from another activity. This attention (or absorption) process is perhaps the last in the life-cycle of an identifiable package of information. It is the stage in which information is transformed into action. Where the form of that information is badly designed it is typically a very costly stage. Consider, for example, how much effort managers must put into reading the listings of raw data taken from an MRP system. Even when the form of information *is* well designed, its absorption by people is still expensive: think of the relative costs to a firm of buying a copy of the *Financial Times*, and of having each of its managers read it. The second surely predominates.

Information should also have a material benefit. It should have a source of value that makes people think it worthwhile incurring its attendant costs. There isn't much intrinsic value to information, so it is reasonable to confine our attention to its application. (Some people *do* find knowledge intrinsically pleasurable, but this is a kind of pleasure which does not figure highly in the operations of a business.) It is the action that follows the acquisition and processing of information, an action that would have been different in some way had the information been different or simply non-existent, that lends information value.

To give a specific instance, the information embodied in the design of a product derives its value from the revenues that it will earn the company. It may be necessary to transform it to a physical good before it can be sold, but one can still speak of a design as having a value at the point at which it is applied. How great this value proves to be will depend on whether the product is a marketable one, whether it can readily be realized in material form, and whether the design is done in good time.

The benefit attributable to a piece of information going through a particular series of processes is determined by the relationship between the form in which it starts and the form in which it ends the sequence. Given a marketing specification, the value of a design stems from the properties it embodies at the end of the design process, regardless of how that process is carried out. The various stages of this process have a great bearing, of course, on the design that is eventually synthesized. But to work out the benefit of the design you need only look at the design itself, in combination with the specification perhaps. This contrasts with the way in which you would need to go about discovering costs: you would have to look at the costs of every stage of the process, and then somehow aggregate them. This difference in the way that costs and benefits are tested is not especially significant here, but it can occasionally yield some useful insights.

In areas where the styling of information is very important, explicit rules about how it should be formed give clues as to where the benefit comes from. There are, for instance, a number of concepts that are supposed to underlie the creation of financial accounts: objectivity, fairness, consistency, materiality, and prudence perhaps. This suggests that information is beneficial to the extent that it is objective, fair, consistent and so on. Such rules seem to be less common when it comes to assessing engineering drawings, NC programs, MRP schedules and so forth. And this is perhaps fortunate because it is usually very difficult to answer such questions as whether objectivity precedes materiality.

The third thing that we know about information is that it is almost always associated with a degree of uncertainty. We can rarely be absolutely confident that a particular piece of information is completely applicable to what we are doing. Uncertainty has many sources: inaccuracy or a lack of resolution during its acquisition is one. Errors and randomness during processing is another. The risk of obsolescence is yet another. A process plan, for example, might suffer from inaccuracy during its creation if the process planner is a poor typist, or has too little knowledge of his firm's production plant. It might suffer an unacceptable lack of resolution if the planner is unable to predict operation times with any dependability. It might suffer errors when it is subsequently read by a foreman or machine operator on the shop floor. It might suffer obsolescence if the company operates inadequate change control procedures.

Uncertainty isn't necessarily inadvertent, for, if the benefits of resolving it fall short of the costs of doing so, it may be perfectly acceptable to retain it. Most of the time, however, it can be thought of as something which reduces the benefit of information and may increase its cost.

Obsolescence is related to a further property – that of timeliness. The longer the delay between the event that first gives rise to information, and the application of that information to a useful activity, the less it is generally worth. The benefit that the information carries is, in other words, discounted to a degree that matches the scale of the delay. As with most things it isn't possible to be categorical about this; for if a piece of information is delayed sufficiently it might, in conjunction with a later piece of information, yield more value than it would if it had been acted on immediately. But information that improves with age is rare. It is normally quite sensible to suppose that the *expected* value of information will decay with age. Age increases the likelihood that it will no longer be synchronized with other pieces of information, that it will be unmarketable, that it will have been compiled under different assumptions from those currently in use, and so on.

Figure 3.3 summarizes the most important properties of information and information-processing chains.

A slightly different way of looking at an information-processing activity is in terms of states (rather than inputs) and actions or decisions (rather than outputs). Information systems interpret the state of relevant aspects of the world and cause a suitable action to be performed. The main function of these systems is therefore

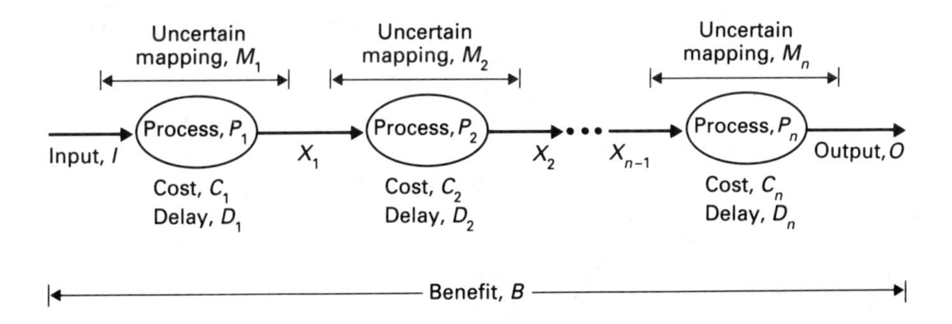

Figure 3.3 Interesting information properties

to distinguish between different states and to allocate to each distinct type a certain result. For example, a control system might operate an actuator if it determines that the state of a production process lies outside certain limits. A decision support system might transmit messages to its users if it detects potential liquidity problems. The role of the information system is in other words discriminatory – it determines when a process is and when it isn't operating satisfactorily.

Flows and structures

In practice, it is rarely possible to consider isolated pieces of information. It is more appropriate to be concerned with the generation and movement of streams of information from one point to another, and with the structures in which different pieces of information are interrelated. Real activities deal with the continual processing of information, with its aggregation and with its classification.

To speak about the *flow* of information is an important way of directing the focus of an organization's attention to its use and value. Looking at flows helps people concentrate on the need to deliver information, and it helps emphasize the point that its benefit becomes evident only when it is applied. This is much like the logistics of managing materials on the shop floor, in which the prominent objective is to achieve throughput – to deliver a stream of products that will earn revenues. Shop floor automation largely came about with the recognition a long time ago of the importance of these product flows,[7] and we have now come to think of the traditional emphasis on keeping work-centres busy by piling up inputs in front of them as being misplaced.

Information accumulated at some intermediate point is, like a work-in-process inventory, simply a drain on the resources needed to store it. It becomes obsolescent, and it hides problems in some of the information-processing functions that are best uncovered. Accumulated information is also, of course, a reflection of processing times. The bigger the accumulation, the longer the lead-time associated with delivering that information, the longer the wait before the investment made in it pays off, and the greater the risk that there will be no pay-off. A big backlog of bills-of-material awaiting entry into the MRP system represents a delayed product introduction or delayed product delivery every bit as much as piles of stock in front of a production process. This affects quality as well – it means that mistakes and shoddiness are only discovered after a considerable time. It may take weeks to find out that our bills-of-material inadvertently call for components of which the MRP system has no knowledge.

It would be nice to be able to manage information in the same way that factories are beginning to manage physical materials and production resources. One would look for ways of allowing sequential operations to be done with the maximum degree of concurrency, and for some means of synchronizing the activity of each operation. Unfortunately, the problem of managing information is even more complex than that of managing materials. Useful materials tend to flow one way – from a receiving point towards a kitting or production process. Information, on the other hand, flows in all directions. If an engineer is to design a part's geometry well, for instance, he will need specifications, standards and catalogues containing information about materials and components, process knowledge and some information about the particular machine tools his company happens to operate at the time. Administrative instructions, advice and acknowledgements are exchanged when the design starts, when it is complete, when it needs to be modified, and so on. It is difficult to see how you could superimpose a flow on such an involved process.

Moreover, many of the points at which information is exchanged between people are not the simple actions that they might sometimes appear. Product designers, for example, don't passively receive information on manufacturability from production engineers and factory foremen. Together, these people negotiate a trade-off between elements that make a product function particularly well and perhaps look good, with elements that make it easy to fabricate or assemble. The act of exchanging information is as much a process of reaching consensus and taking a substantial interest in other people's difficulties, as it is of performing an act of administration.

What you *can* do to some extent is to take an abstract view of information flows by ignoring most of the detail. Just as people worry about the minutiae of booking operations after the basic ideas of shop floor logistics have been worked out, so they can postpone worrying about the precise protocols to which they will subject information processes. You can also draw a box round the main types of information process to exclude those that cover what are essentially support functions. You would need to exercise some care here, of course, because it is important to understand the operation of the information system as a whole, rather than the separate operation of its components. Nonetheless, it is reasonable to exclude activities like accounting and computing whose outputs do not earn the company revenues. It makes sense in other words to concentrate on the sequence of value-adding activities that most firms operate – the activities that materially affect the form of the final product and that, as a result, determine the revenues that a firm collects when its product is sold.

A discussion of how a model of this sort might be prepared in practice may be found elsewhere. The important point for the purpose of this chapter is that one of the main views that we will want to take of a technology will record its impact on information flows. We might ask whether it improves their

- cost,
- accuracy,
- volume, and
- speed.

Greater accuracies will be most needed when a firm is attempting to improve the quality of its products. The ability to manage greater volumes will be important when a firm wants to increase the variety of its products, perhaps to tailor them more closely to its customers' needs. And achieving greater speed in information flows will be especially necessary when a firm wants to reduce its lead-times, perhaps to bring products more quickly to market.

It is an interesting point that this division into four basic properties mirrors the characteristics we would expect of an organization conforming to the demand model of Chapter 2. This suggests a way of identifying which of the four properties might be the most important at a particular time.

To look at information flows is not, of course, enough, even at the early stage in which a firm is evaluating a proposal for new technology. The other main aspect of information is its structure – how large pieces are made up of small pieces, what the pieces mean and the formats in which they happen to appear.

Information structures tend to be quite stable over time, especially at the more general levels that record the interrelationships of complex objects. The implication is that to understand how best to plan the mechanisms for recording and manipulating information, the focus should be on these stable structures. They will provide the basis for dividing up the information management task between different systems. In showing how one piece of information is related to another they also provide a kind of map showing people the information that is available within the organization and the significance that it has. They should make it obvious, for instance, whenever two objects contain the same information: this might inadvertently become inconsistent unless the necessary controls are added.

But, although in technological terms information flows and information structures deserve to be treated with a degree of equality, it is the flows that are the predominant influence on economic issues. Little more will therefore be said of structures.

3.3 Classifying a technology's benefits

The purpose of this section is to discuss the manner in which the benefits of a new technology can be classified, and to suggest how several common types of application might fit into this classification.

The point of building up a classification of benefits, rather than simply compiling a series of lists, is to maintain a feeling of wholeness about the evaluation process. We know from the previous section to look for the impact of new systems on the various properties of information flows, and we know to look at the impact on both effectiveness and efficiency. We do not, so far, have a way of fitting these different aspects together. Another reason to build a classification, even if it is a rudimentary one, is that it becomes possible to speak about issues at different levels of detail without, at the same time, becoming too vague or irrelevant. It might be appropriate, for instance to discuss the scale of commercial costs in general before exploring the scale of particular types of commercial cost – such as the cost of information gathering and the cost of carrying out transactions.

A classification helps resolve some of the fuzzier ideas into elements that are more immediate and easier to grasp. This makes it more realistic to consider quantifying the effects of such elements. With a structure for ordering our thoughts about technological benefits, we also have a scheme into which it ought to be feasible to slot in new applications that we have not so far come across.

A perennial problem with classification schemes is that they become obsolete if they are not changed to reflect new interests and new discoveries. That was the point of saying in the preamble to this chapter that the classification that follows is only a starting point for understanding the effects of technology. So rather than suggest that the classification is the right one (or even that it is a complete one), it is more sensible to say that it is of the right *type*. Classifications also rely on a

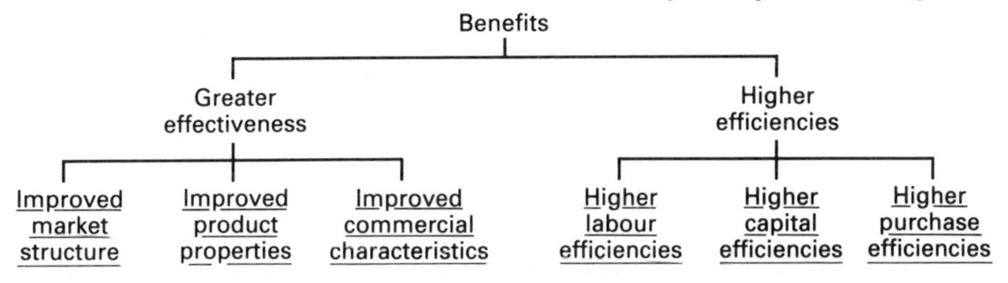

Figure 3.4 Benefits classification (1)

certain amount of intuition in the way they draw boundaries between one issue and another: sometimes the divisions are obvious ones, but sometimes they are rather arbitrary. The naturalness with which they summarize the issues involved depends heavily on local circumstances, and there is really no substitute for the technologists and managers of individual firms coming to their own understanding of what might be of benefit in their organizations.

A classification

Figure 3.4 is the highest level of the benefits classification. It indicates that the most basic characterization of a benefit is whether it is concerned with greater effectiveness or greater efficiency. (Any one development can affect both of course, in which case it has more than one benefit.) At a slightly more detailed level, we can speak of improvements in effectiveness as being those due to a better market structure, better product properties or better commercial characteristics. To see what distinguishes product properties and commercial characteristics in these terms, you would have to go a further level down: the fact that these classes are underlined indicates that they are detailed on another figure. The essential distinction, however, is that product properties are the technical attributes embodied in the physical form of the product, while commercial characteristics are elements such as price that are assigned to the product when it becomes a marketed good.

Higher efficiencies are somewhat arbitrarily divided among labour, capital and purchasing effects. These reflect the firm's ability to make better use of its people, of its earning power, and of the materials, energy and services it buys from elsewhere.

The remaining figures look at each of the lowest classes in Figure 3.4. Figure 3.5, for instance, indicates how we might classify favourable developments in market structure. Raising entry costs for firms not currently in our market, but considering entering it, is one such development: this in turn might stem from exploring economies of scope or economies of scale (if they exist). If there are such economies, any firm that cannot match the variety or size of an existing participant's output will have a higher cost base. Reflecting the discussion of the previous chapter, a further level of detail is added to the classification of scope economies, suggesting that they are either instances of improved engineering re-use, or of greater flexibility in any of the firm's processes.

The heading of greater *re-use* is intended to suggest the possibility of avoiding engineering activity by applying the products of earlier work. Technology can be of considerable assistance here, by helping people to find, understand, modify and compare existing designs. Some electronic document control schemes allow their users to establish classifications, while others let them use key words. Hypertext is also a promising technology for re-use: it can explicitly record the associations between one document and another that can be used to understand how the documents had earlier been applied. And in a number of areas it is possible to use parametric programs – schemes in which one can obtain any number of outputs by simply changing the values of a number of scalar inputs. For instance, a designer might re-use a basic part geometry to produce different drawings by supplying a program with different dimensions, quantities, materials and so forth.

As usual, however, the technology is not enough on its own, for people have to work in a way that supports the idea of re-use. This means investing in a certain amount of work to re-classify existing designs, and often to modify them – to make them slightly more general. And when it comes to the process of creating new designs, the idea of making them general enough to be used elsewhere tends to contradict the management controls typically applied in engineering companies. There the stress is laid on minimizing the costs of contracts currently underway, with little regard for subsequent work.

Another type of improvement in market structure for existing suppliers is a lower rate of customer turnover, and this can sometimes be encouraged by increasing the opportunity costs of unwinding established relationships: either by raising the price of switching from one supplier to another, or by lowering the price of existing transactions. These transaction costs are the same as those discussed in the previous chapter. As also suggested in Chapter 2, computer-based systems can improve the broking process. Electronic technology offers fast and easy access to large amounts of easily-updated details about different products, and it becomes feasible to have a central source of comparative information. This reduces both the marketing costs of suppliers and the shopping costs of customers.

Figure 3.6 proposes some types of benefit that would be classified as being improvements in product properties. As far as manufacturing technology is concerned, the properties of most interest are those such as variety, fitness for purpose and innovation. Both variety and innovation tend to depend on similar aspects of the firm's process – the time it takes to bring new products to market,

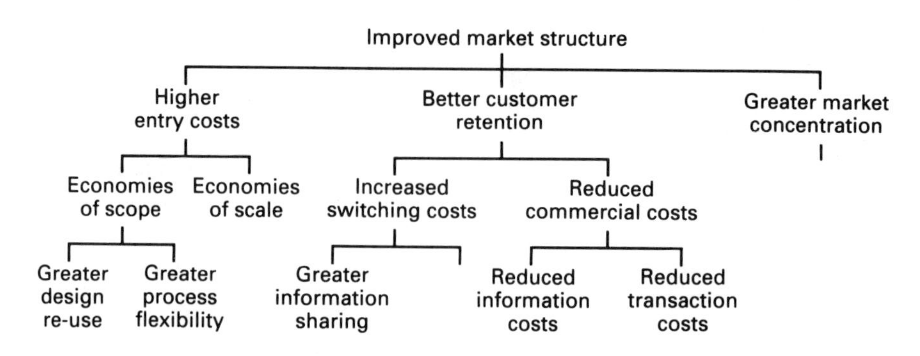

Figure 3.5 Benefits classification (2)

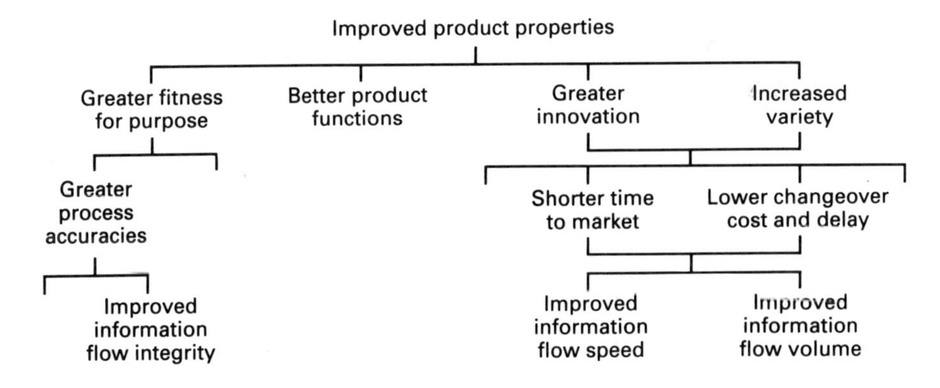

Figure 3.6 Benefits classification (3)

and the cost of switching operations between one product and another. These in turn are determined very much by the success with which the firm manages its information, particularly its ability to accelerate information flows and increase their volume. Faster flows are needed for shorter lead-times, and wider flows, as it were, for greater product variety. Improvements in fitness for purpose, on the other hand, depend especially on greater information flow integrity – on reducing the number of errors made in the production and interpretation of information, perhaps.

One can think of a number of ways of achieving these benefits: by implementing a sophisticated, rule-based validation of data as it is generated, by automating the transcription of data between one application and another, and by improving the presentation of data to its human users.

Figure 3.7 shows how the class of improved commercial characteristics might be subdivided. Lower product prices are often made possible by lower costs, and therefore by higher efficiencies. Improvements in delivery performance are improvements either in dependability (fulfilling delivery promises more regularly), or in the length of delivery periods. Both can be achieved with better process accuracies. By making sure that processes of any sort meet the specifications transmitted to them, the need to rectify and scrap products is avoided. And the previous figure suggested that such accuracies may be obtained by improving the integrity of information flows.

There are other ways of achieving shorter delivery periods of course, all potentially within the scope of newer technologies. Some of these are directed at reducing the time taken to perform individual processes, while others are intended to reduce the elapsed time associated with a complete pattern of processes – by increasing the extent to which they can operate concurrently, for example. Systems that record and apply the precedence between one information-producing activity and another can ensure that there is no unnecessary delay in the organization's information flows. This is one of the benefits of what are sometimes called procedural automation schemes.

Figure 3.8 is a classification of the first type of efficiency – that associated with labour. Issues such as greater focus and drive lie mostly outside the technological

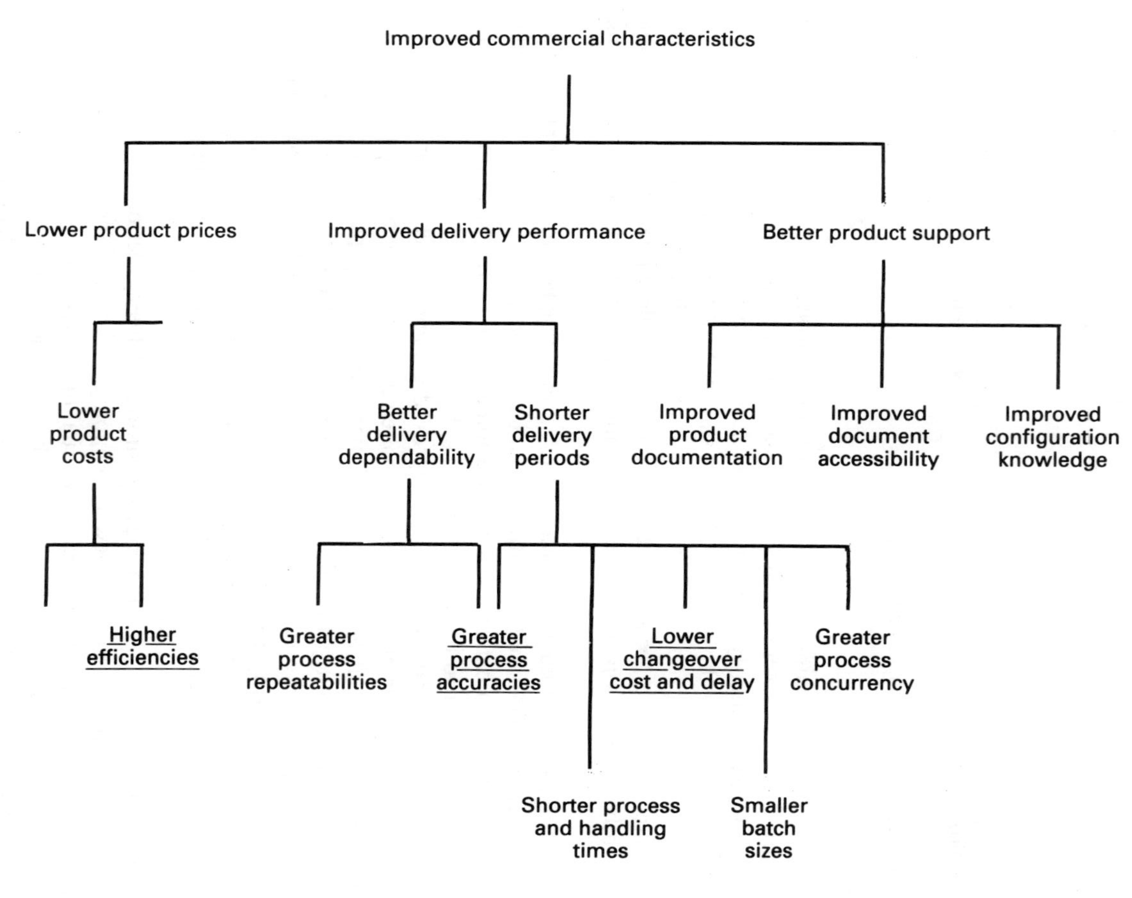

Figure 3.7 Benefits classification (4)

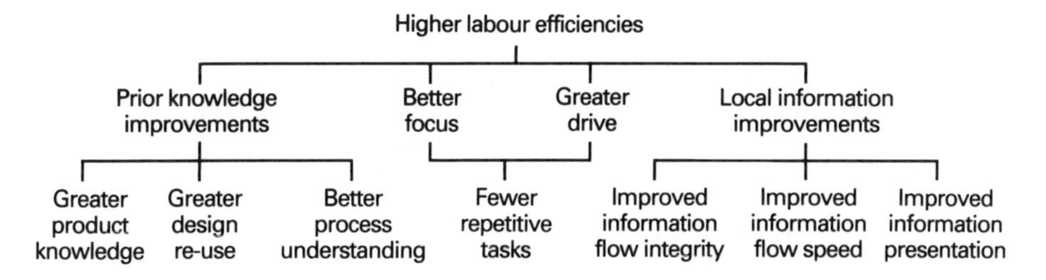

Figure 3.8 Benefits classification (5)

domain, but the use of technology is sometimes an appropriate way of relieving people of their more dispiriting jobs. The main contribution that can be made to human productivity is to improve the information that people act on as they work, whether it is the prior knowledge that can be expressed in terms of rules and procedures, or the local knowledge that is usually cast in terms of facts and data.

Better prior knowledge becomes available when they can re-use the expertise and experience of others (perhaps by re-using product designs), and when they can apply a greater understanding of processes (perhaps the diagnosis of machine faults). Local knowledge can be improved by making people aware of facts more quickly, more dependably, and in a way that properly captures their attention.

To a limited extent, technology can be used in place of people, and at a lower cost. But this is not a central rationale for most new systems, and it is not an experience that is typical in firms that have introduced new systems. It is really quite fortunate that the distinctive characteristics of computer programs – the speed at which they can be run, their repeatability, literalness and ease of duplication – are so different from the things that make humans human – their self-awareness, intuitiveness and ability to generalize. Because the two sets of characteristics do not often compete we can, potentially, get the best of both worlds.

In Figure 3.9 the meaning of capital is simply that it is anything accumulated in order to conduct manufacturing activities in the future. It therefore includes managerial know-how as well as finance, plant and machinery. It is perhaps a rather arbitrary decision to put managers' productivity here rather than in Figure 3.8, but here management is meant to suggest anything that plays a role in determining, directing or modifying a firm's operations, rather than being a direct

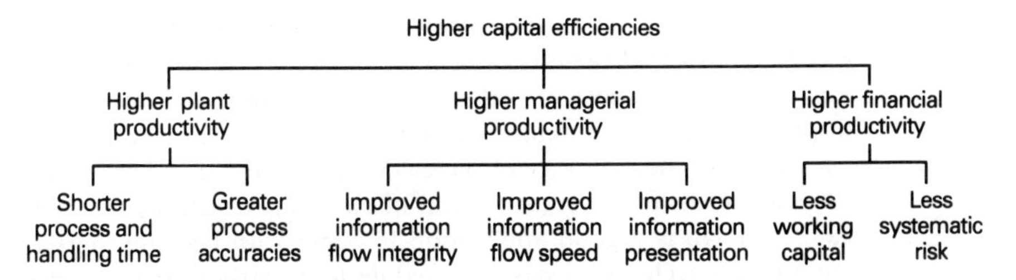

Figure 3.9 Benefits classification (6)

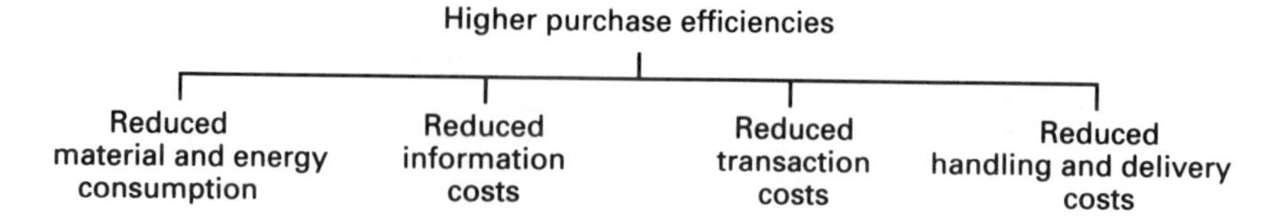

Figure 3.10 Benefits classification (7)

part of them. It relies of course on receiving information effectively.

The ability to reduce the risk in the firm's performance is shown as an improvement in financial productivity. The exact meaning of this, and the importance of its impact, will become evident in Chapter 5. The essence, however, is that the better a firm can ride out fluctuations in its environment, the better the chances that the returns to its financiers will meet their expectations, and the lower the price they will charge for their funds. The flexibility to meet rapidly changing customer demands, for instance, might reduce a firm's funding costs.

Savings in working capital have been placed under higher financial efficiencies because they are essentially concerned with reducing stocks and flows of the more liquid types of asset. A fall in work-in-process inventories, for instance, yields a one-off benefit when the surplus is liquidated, and a recurring benefit from the saving in holding costs: obsolescence, storage, insurance and so on. It is also a common observation that less work-in-process is associated with shorter lead-times and better quality, although big inventories are more a symptom of an incorrect approach to organizing the production process than they are a cause of it.

The final part of the classification (Figure 3.10) describes the efficiency associated with obtaining things from other firms. This partly concerns the buying process, so reductions in the cost of obtaining information about available commodities and services, and in the cost of performing commercial transactions, will improve this efficiency. (We might look for the benefits of commercial electronic data interchange in this area of the classification.) It also concerns the ability to reduce the need for such things as materials and energy, and to improve the manner in which they are physically delivered and handled within the factory.

Some applications

We can now look at a couple of specific types of system in the light of this classification. The process is simply to link a system with appropriate benefit classes in the hierarchy, and then to explore the ramifications by looking at the components of the hierarchy that surround these classes. If we could say, for example, that a given technology reduced the cost of making commercial transactions it is easy to see that this promotes greater purchase efficiencies. We might, however, wonder whether this is the most effective way of increasing efficiency; whether it would not be better to reduce the firm's consumption of materials or energy.

Computer-aided design is a good application to begin with because there is a common feeling that the technology has not lived up to its early promise, perhaps because its backers were looking for types of benefit that it could not provide.

Since CAD is very much a tool for continual and direct human use, we might start by looking at the ways in which it can improve human productivity. To a limited extent it might reduce clerical tasks (Figure 3.8) by removing some of the more mundane jobs associated with drawing on paper – messing about with drawing sheets, dyeline reproduction and so forth. To some people, however, it is more pleasant to work with tangible materials rather than electronics, and the savings in people's time are usually marginal.

It is more likely that benefits will become apparent when there is scope to re-use designs, because the malleability of something in an electronic form is then much more important. With CAD it is at the very least much easier to erase old elements and insert new ones when preparing a revision or a variant of an existing design. We can therefore plan for benefits associated with greater design re-use (Figure 3.8). And we might attempt to magnify the CAD system's potential for re-use by adding additional mechanisms, such as cataloguing and classification schemes, that encourage people to think in terms of re-using existing work much more often. It is also apparent from Figure 3.5 that re-use contributes towards scope economies, as well as human efficiency. This means that a company that makes proper use of CAD can engineer a variety of products – perhaps using a limited range of components – that a company using manual drafting methods could not.

Figure 3.5 further suggests that a firm could achieve commercial benefits by reducing the costs of exchanging information with its customers, and that it might as a result be worth exploring the possibilities of exchanging drawings in electronic form. Once an inexpensive and effective channel has been established between supplier and customer, each has an incentive – other things being equal – to maintain the relationship. A similar line of argument can be applied to information flows within a manufacturing company. It is generally cheaper and quicker for one person to build upon another's work if he can manipulate the latter directly, rather than re-draft it. He might even be able to use tools that automatically transform designs: from schematics to net-lists or physical layouts, for instance. Any point in our hierarchy that points to improvements in information flows could therefore be applicable: we might plan for shorter times to market (Figure 3.6), or greater human efficiencies from better local knowledge (3.8).

In these respects, CAD has much in common with other design tools (such as computer-aided software engineering), and with other tools used for document preparation (such as desktop publishing). Perhaps the most important element of each is that it is not the process in which one begins a design from scratch that is the most affected. It is the process in which people engineer things by reproducing, varying and combining existing designs. Since much of the knowledge that underlies any document is contained in its predecessors, this plainly ought to be an important process. Software tools should make engineering cheaper in the sense that one doesn't have to repeat the process of converting knowledge into a form that can be sold or transformed into physical products, and they should help a firm to avoid losing the underlying knowledge itself.

Some of these systems – notably desk-top publishing – have an obvious effect on the presentation of the information they are used to produce (Figures 3.8 and 3.9). Effective presentation can direct the reader's attention to just those contents that deserve the highest priority, and it can in certain cases make the reader better disposed towards the writer. It is sometimes the case that such systems improve the quality of documents, in the sense that they embody a greater amount of information with fewer inaccuracies. But it is probably evident to most people that such systems are not always used with the most tasteful results, and that prettiness is less of an issue than clarity, completeness and accuracy.

A technology that is generally applied close to the opposite end of the manufacturing process is that of robotics. Robots are often used to replace people (a loss of flexibility, perhaps, in the expectations of a gain in repeatability), and to replace dedicated handling equipment (a gain in flexibility, perhaps, with a loss in speed). Where they stand in for dedicated equipment, robots are likely to be a source of scope economies (Figure 3.5): they represent a resource that is sharable between the production processes of different product lines. They are also a mechanism for improving information flows – not in the sense that they deliver information more effectively, but in the sense that they can *apply* it more effectively. Robots are some of the more advanced mechanisms we have for converting information into physical work. This is especially so when there is a quick and accurate way of transforming product information to production information. This transformation might take the rather rudimentary form of using simple product parameters, such as physical dimensions, in order to make choices in a fixed part program. It might equally take the form of a process invoking different part programs according to the product type being manipulated.

The ability to use robots genuinely as flexible resources will also make product changeovers quicker, and perhaps reduce the time taken to bring new products to market (Figures 3.6 and 3.7). Properly chosen, robots should also improve the dependability of assembly processes by providing greater accuracies and repeatabilities (Figures 3.7 and 3.9) – particularly for products that are difficult for people to manipulate, and in harsh environments.

As usual, the economies that robots promise take some thought and effort to realize. Programmers have to strike a balance between the cost of programming a particular application, and the benefit of building programs that are general enough to be re-usable. They also have to devote some effort to compiling and managing program catalogues. The system's engineers need to pay a good deal of attention to the channels and protocols that will be used to get information to the robot's controller, and to the formats it will have to be presented in if it is to be interpreted correctly. They will need to be careful that they create the conditions in which the robot's accuracy of movement can be obtained in the first place, and then maintained over extended periods.

They have, in other words, actively to seek the benefits of a technology if there is to be any likelihood of experiencing them.

3.4 Summary

Computer-based manufacturing systems have a number of characteristics that distinguish them from older types of plant. They are, for instance:

- more flexible in their operation, since they can be programmed to take large numbers of decisions;

- more malleable in their construction, since software is much less costly to change than hardware of a comparable complexity;
- capable of abstraction – of presenting informative views of the processes they control.

The way in which their introduction is managed is made harder, however, by a number of further characteristics:

- their novelty, in their internal form but more significantly in their external operation;
- their reach – the fact that with an ability to manipulate information they have a far greater scope than mechanical systems:
- their potential for further development, which means that it is hard to conduct a definite appraisal of their effects in advance;
- the extended nature of the chain of causes and effects that connects a technology's application with its financial consequences.

Since these systems are primarily information processors (rather than material processors), it makes sense to understand the logistics of information – in a manner similar to that in which it is common to think about the logistics of materials on the shop floor. The most important idea is that of information flows, since these emphasize the need to deliver information in order to extract some value from it. Stocks of information are valueless until they are converted into flows. The most relevant properties of these flows (and of the processes between which they are channelled) are perhaps

- their costs,
- their delays,
- the uncertainties in the various transformations they undergo, and
- the benefits gained on their conversion into actions.

Finally, the chapter closed with a discussion of how such benefits might be classified. The intention was to illustrate a process in which one can form a reasonably systematic view of why advanced technology is worth adopting. The classification illustrates how increasingly specialized (and more concrete) types of benefit can be identified. Equally, it suggests a way of generalizing on specific kinds of benefit so that an understanding of the value of one system can be carried across to others.

Notes and references

1 Simon, H. A. Decision making as an economic resource. In Simon, H. A. *Models of Bounded Rationality. Volume 2: Behavioural Economics and Business Organization* MIT Press, Cambridge (Mass.) pp. 84–108 (1982)

2 Boddy, D. and Buchanan, D. A. *Managing New Technology*, Basil Blackwell, Oxford p. 25 (1986)

3 Gerwin, D. and Tarondeau, J. C. Case studies of computer integrated manufacturing systems: a view of uncertainty and innovation processes. *Journal of Operations Management,* **2**(2), 87–99 (1982)
4 See, for instance, Ferguson, P. R. *Industrial Economics: Issues and Perspectives*, Macmillan Education, Basingstoke (1988)
5 Banker, R. D., Kauffman, R. J. and Morey, R. C. Measuring gains in operational efficiency from information technology: a study of the Positran deployment at Hardee's Inc. *Journal of Management Information Systems*, **7**(2), 29–54 (1990)
6 Simon, H. A. The impact of new information-processing technology. In Simon, H. A. *Models of Bounded Rationality* pp. 109–33
7 Einzig, P. *The Economic Consequences of Automation*, Secker and Warburg, London p. 30 (1957)

Part II The Appraisal of New Systems

4 Some informal approaches

I can stand brute force, but brute reason is quite unreasonable. There is something unfair about its use. It is hitting below the intellect.

Oscar Wilde *The Picture of Dorian Grey*

The purpose of this chapter is to begin to look at how a firm can assess the worth of a specific investment in new technology. Most of the underlying ideas are relevant to any type of investment in a manufacturing firm, and many have an applicability that extends even beyond commercial organizations. Since most firms attempt to get the greatest possible benefits from the resources at their command, and since they would mostly agree on what is beneficial and what is costly, this is hardly surprising. It is largely in the way that these ideas are applied in practice that new technologies demonstrate special characteristics.

This chapter is concerned with some of the simpler approaches to investment appraisal, and their simplicity alone means that they have a good deal going for them. It probably means that they are quick to grasp and quick to apply. But it also suggests that they offer no way of dealing with much of the complication presented by real systems and real organizations. And these approaches tend to make unwarranted assumptions either about the firm carrying out the appraisal or the scheme to which it is being applied. The least we can do is to identify what the limitations are, and to anticipate the circumstances in which they are likely to be exceeded.

4.1 Desirable elements

It is easier to say what are, and what are not, the desirable properties of an appraisal system once a few examples have been reviewed. In particular, their various drawbacks usually become evident after a while, and on the basis of these it is often reasonable to generalize about whole classes of appraisal system. The difficulties, for example, associated with methods that yield widely varying answers according to the person applying them suggest that one desirable element is a system of values that can be tested independently of one person's opinions. But to make sure that the right questions are asked, it is worth discussing very briefly the more important properties that we would expect an appraisal system to exhibit.

We first need a way of setting the benefits expected from an investment against the sacrifices that it will demand. This will probably mean reducing each of the benefits and each of the sacrifices to a common scale, and where it is most natural to express them by different scales there will have to be a method of translating quantities between them. It would be difficult to compare, say, the advantage of greater customer satisfaction with the disadvantage of disrupting production, without going through an intermediate stage of reasoning – a stage in which we decided that a certain amount of one would be equivalent to a certain degree of another. Once the various effects of an investment have been expressed in a comparable form, a judgement can be made of the investment's net worth.

Since the appraisal process is used to compare the relative merits of alternative courses of action, our common scale needs to be one that can be applied to any type of investment. It shouldn't misrepresent important effects for spurious reasons – it shouldn't, for instance, exaggerate or downplay any of the benefits of a new technology. And because the providers of a firm's funding will want to compare the firm's opportunities with those elsewhere, the scale ought also to be common to them. This suggests of course that we use some kind of monetary yardstick. We ought nonetheless, to be wary of trying to compare things that are simply incomparable. If no amount of monetary profit could compensate for diminished levels of human safety then the two should not be forced to lie along a common scale.

This common scale defines not only a unit of expression, but also some rules of measurement. These rules are very simple. The most prominent of them is that the costs of alternative courses of action are taken to be *opportunity costs*. This simply means that we regard the sacrifice we make in pursuing a particular action (that is, its cost) as being the value of the next best alternative, which we have to forego. If a programmer can earn his employer revenues of £1000 by working on development X, then the opportunity cost of foregoing X in order that he may work on development Y would be £1000. The cost of Y is therefore £1000.

When resources are bought explicitly the opportunity cost is easy to spot: it is simply the purchase price. In other cases, we have to consider alternative uses of a resource in order to find its opportunity cost. This kind of reasoning is in fact quite natural, but it tends to be obscured by accounting conventions. If one thinks about costs only in terms of charges written off against a profit-and-loss statement one can easily miss the sacrifice that is really made when employing a resource in a particular way.

Our second need is for a way of making explicit just how uncertain we feel about the prospects for a new project. We would also like a way of saying how important this uncertainty is when set beside the project's costs and benefits. Most people and groups are averse to uncertainty, and, all other things being equal, they would take a course of action whose outcome could be predicted with confidence rather than one surrounded by doubt. Both the extent of this doubt, and the impact of an unfavourable outcome taking place, should be registered during the appraisal process. We might want to discount the net worth of an investment to the extent that it embodies uncertainty, or we might want to gather more information in order to reduce it. If we can put a price on uncertainty, we have a way of knowing how much we can sensibly spend on this additional information gathering.

A third need is for a system that distinguishes clearly between objective knowledge about a proposal and subjective judgements about its worth. It is not, of course, reasonable to think that a way of making decisions in a complicated and changing world can manage entirely without calling on people to make judgements. But where it can be done, it makes sense to use information and measures of value that groups of people can all observe and all agree upon. It is usually straightforward, for instance, to place a value on things bought and sold in markets, for they have a visible price.

This idea of using market prices can be extended to the system of values by which investments are assessed. We might want to say that an appraisal system which is based on the way that financial markets value assets is an appropriate one for a firm whose overriding purpose is to maximize the benefit due to its shareholders. Since the operation of financial markets is open to a good deal of scrutiny, it is probably possible to work out such a system of values in a largely objective fashion. This might tell us, for instance, the premium we should demand in the returns of investments that carry a certain amount of risk. In other words, in place of a method in which the individual decision maker determines how much risk he, personally, is prepared to trade for an extra £1 of benefit, we could attempt to discover how the finance market as a whole would price risk.

A fourth need is for stability and consistency. The appraisal process should be such that it yields roughly the same results whenever it is applied to the same set of information. If the results varied sharply from one time to another, according to the mood of the people involved in the process, they would be much less convincing as a basis on which to take decisions that committed a firm to a great deal of effort over a long period. If the results depended too much on the analyst that derived them, this would again undermine their usefulness in an organization that normally outlasts specific individuals. The raw estimates that are fed into an appraisal model are bound to be subjective, and they are bound to be affected by the experience and temperament of the person who supplies them. But we do not, at the same time, want the structure into which they are slotted to be dependent on whims. The rules of the appraisal process, if you like, should be explicit, reasonable and generally accepted by those who will act on them.

A highly desirable property of an appraisal system is that it should help decision makers to use the past as a guide to the future whenever it is sensible to do so. This suggests that it should be based on a set of principles similar to those that underlie reporting systems. This is rather an onerous requirement because it implies not only that the yardsticks should be common to both appraisal and reporting, but that the yardsticks should be applied to the same partition of the organization. In other words, their scope of application should include exactly the same people, projects, departments and so on. In practice, of course, financial accounting systems are rarely consistent with appraisal systems: they are based on concepts such as accrual which are incompatible with the idea of opportunity costs, and they typically summarize results within specific areas of the organizational structure. They are not normally applied to identifiable projects, such as new computer systems, and they will be especially unhelpful when those project have effects that cut across convenient organizational boundaries. We know that such effects are characteristic of computer technologies.

Learning from past experiences has a number of other problems. It is rarely possible to distinguish correlation from causation – to say, because a firm's profitability improved just after a new system was introduced, that it was the new system that caused the improvement. Most firms that want to survive are pursuing a number of developments at any one time, and they do so within an environment that is continually changing. It is hard to disentangle the effects of such changes from each other.

There is, of course, the fallacy that people commonly apply whenever they look back to the results of their own decisions and their own efforts: if an outcome appears to be favourable, they will attribute the success to their own abilities, whereas, if it appears not to be favourable, they will normally connect it with external factors over which they have no control. And there is a sense in which a slavish adherence to history discourages active management in the present. It is very tempting for those resisting change to point to problems in the past and to say that they will be repeated if the company tries the same kind of development again. In reality, managers have a good deal of scope for creating conditions in which innovation is successful, and failure in the past may provide the information needed for success in the future. The introduction of flexible manufacturing systems is a case in point. Some companies have expressed a certain amount of regret about adopting FMSs because of what turned out to be their poor returns. They have perhaps lost sight of the extent of the experience they gained, and how this might be applied to take proper advantage of such systems in the future.

Lastly, the appraisal process falls within the normal operations of a firm – it doesn't stand above them. This suggests that the process itself should satisfy the basic test that its benefits sustainably outweigh its costs: that the decisions based on it are so much better than those that would otherwise have been made that they justify the costs associated with gathering and processing information, making forecasts, reaching consensus and so forth. The benefit of the appraisal process is, unfortunately, not an easy thing to measure, particularly as one can rarely say anything about the worth of a course of action that the process rejected. However, in the absence of other deciding issues, we should plainly look for appraisal systems that are cheap, quick and comprehensible to the people applying them.

4.2 Scoring models

A glance at the contents list at the beginning of the book would indicate that the informal approaches considered here fall under one of two main headings – scoring models and rules of thumb. Scoring models essentially provide a way of saying that an investment is good in some respects, poor in others, and that there is a certain residue when these elements are placed beside one another. Rules of thumb are based on yardsticks like payback period which may not measure an investment's value directly, but which as a matter of experience tend to be associated with it. The more formal principles described in later chapters are about going back to the basic financial processes that take place in a developing industrial firm. From these principles follow some powerful and attractive methods for assessing the worth of such developments as computer-based systems.

Each group of methods has its advantages and its drawbacks, but for most applications there are overriding reasons to use financial principles. In fact, the main point of discussing scoring models and rules of thumb is to show why, when they are applied to the evaluation of advanced developments, they are not entirely up to the job.

Basic ideas about scoring

How much does an investment in, say, a process planning system reduce the time needed to introduce a new product? What is the impact of this reduction in lead-time on the flexibility of the factory and on the size of its order book? How important, in any case, are flexibility and turnover? How important are they when set against an increase in the factory's fixed expenses? What can be concluded – should the company introduce such a system or not? These questions perhaps represent a simplification of what tends to be asked in a real situation, but they are questions that suggest the need for a scoring model of some kind.

The basis of such a model is very straightforward. First, a series of interesting properties needs to be identified, and their relative importance expressed by assigning them weights. The alternative developments (or courses of action) are then rated against each of these properties, and the score of each alternative is found by adding the products of associated weights and ratings. The alternative with the highest score is supposed to be the one then adopted. In other words.

$$\text{Score of option}_i = \sum_{j \in \text{all properties}} (\text{rating of option}_i \text{ on property}_j \times \text{weight of property}_j)$$

Approaches such as this are usually described as being linear, compensatory and subjective. They are linear in the sense that they are just sums or differences of weighted ratings; compensatory because a poor rating on one characteristic can be offset by a high rating on another; and subjective in that the rating and weights are obtained by judgement rather than by a statistical analysis of past decisions.

The properties used for rating new developments reflect the issues of most importance to those making the decisions. They might be indicators of financial performance – such as the ratio of profits to assets over a particular period – and indicators of financial position – such as the ratio of liquid assets to short-term liabilities. They might reflect issues of commercial significance in a company's markets: perhaps contract lead-times, new product introduction times, product quality, product price, and the degree of product innovation. They could be indicators of the effectiveness of the firm's internal processes: engineering costs and engineering lead-times, product changeover times in production departments, and so forth.

Such attributes will plainly vary from industry to industry. The issues that are important to contractors will not be the same as those considered important by process industries or producers of consumer durables. Even basic financial properties (such as the liquidity of a company's assets) will have a significance that changes from one industry to another. The attributes will also vary from company to company within the same industry. A particular organization might be having specific problems that, at least in the short term, draw its attention to a few, salient features of its operations – perhaps the unhealthy scale of its work-in-process inventories. And companies will have different areas of comparative advantage. It is natural that a firm will be most interested in the effects of new technologies on such areas.

There *are* nonetheless a few common themes, and some of the results of a survey conducted for CAM-I[1] may be of interest to firms exploring the sorts of issue they should consider as scoring properties. Key factors consistently identified by different businesses include such things as the quality of suppliers' products, the quality of production processes, inventory reduction, the cycle time of product introductions, throughput and manufacturing times.

Once the set of scoring properties has been chosen an analyst has then to work out a scale of relative significance. In other words, each property needs to be weighted. The intention is that if a potential development can influence one of these things for the better without detriment to the others then it will score highly. If it brings benefits to one of the properties but has an unfavourable effect on another, then it is the scale of the impact on each property in combination with the properties' relative importance that determines whether on balance the development is worth pursuing.

The weights can be established by stating one, well-defined objective, and by then assessing the degree to which each of the properties contributes towards its achievement. Given, for instance, a firm that has the single objective of maximizing its return on fixed assets, its managers might decide that shorter lead-times are twice as important in achieving this as everything else put together. They might assign weights (arbitrarily taken as percentages) of

lead-time 50% : everything else 50%

Perhaps 'everything else' can adequately be expressed in terms of better quality and lower price, if there are no other factors of comparable significance. If better quality is one-and-a-half times as important as lower price then the scale of importance works out as

lead-time 50% : quality 30%: price 20%

The lack of any model underlying such figures, and the fact that this set of relative values is quite likely to change over time as the firm's circumstances and capabilities develop, are significant drawbacks of this approach. It fails to capture the reasonably constant economic rationale of the firm: the long-term purpose that lies behind the firm's changing interpretation of how it can achieve a strong performance.

The next step is to apply the model to a specific decision, to rate the subject of the decision according to its impact on each property of interest. Again, it is the relative importance of the impact that is important: to what degree, perhaps, the effect on quality differs from that on price. In rudimentary forms of scoring model (an alternative form will be described later) this impact has to be gauged on a scale such that an increasingly favourable impact produces increasingly positive numbers. For example, lead-time effects might be expressed by the percentage reduction in delivery periods, and quality in terms of the percentage reduction in product faults throughout product lifetimes. A degree of care has to be exercised here, because a linear model implies that a reduction in lead-times of, say, 20% is exactly twice as significant as a 10% reduction. This is unlikely to be the case since most factors tend to be associated with a law of diminishing returns.

These scales also interact with the weightings. By saying that quality is one-and-a-half times as important as price, one is in fact saying that a one per cent reduction in faults is one-and-a-half times as important as a one per cent reduction in price. If the price effect were re-expressed in terms of absolute reductions (such as £x rather than y%) the weightings would need to be changed. The alternative is to speak in terms of abstractions such as degrees of significance, and this, in practice, seems to be the way that models are usually built.

Continuing with our rather sparse example, suppose that an investment in a new system (call it System Z) is being assessed against the alternative of doing nothing. Suppose that the scales for each of the three properties are stated in terms of degrees of significance (in a range of zero to ten), and that the new system is judged to have an impact on the three properties of lead-time, quality and price of 6, 5 and 2.5. Doing nothing would apparently have an impact of 0, 0 and 0, although of course in many situations external pressures would mean a deterioration in the firm's fortunes if it took no action at all. The net scores are, for System Z,

$$50\% \times 6 + 30\% \times 5 + 20\% \times 2.5 = 5$$

and for doing nothing

$$50\% \times 0 + 30\% \times 0 + 20\% \times 0 = 0$$

leaving System Z an obvious winner. In reality, the analyst would hardly be discharging his duties thoroughly if he didn't find additional alternatives to set against the proposal for System Z.

It is sometimes possible to squeeze more out of the results, and more out of the process, of building a scoring model. One might be able to gauge the significance of the disparity between the scores of the different options, and the extent to which they fall short of the maximum possible score. In practice, however, models of any real interest are considerably more complicated than this example suggests. For a start, there may be several layers of scoring property. At the top lie broad, fundamental properties such as the return on investment. At intermediate stages come aggregate properties that are important only because they influence the return on investment – things like the scale of inventories. At the bottom are detailed effects, such as product changeover times in a particular cell. Something at an intermediate or lower level may well affect more than one property at the next higher level.

A couple of examples

A paper by Lehmann[2] describes a scoring system for the appraisal of developments in smaller and medium-sized companies. There are several methods within this system, and two of these are of interest here: one called the rating system, and one called the connectance model.

The rating system is based on weighting issues relevant to a company's performance (like lead-times, quality and flexibility) on a scale of 1 to 10, according to how important they are thought to be. A further value from 1 to 4 is then allocated to each issue to express the confidence with which it has been weighted. This is a useful addition to the basic compensatory model: not only does it encourage the analyst to be explicit about the degree of uncertainty he feels in making his choices, but it also indicates in which areas further information, or further thought, is needed. The extent to which the proposed development affects these performance issues is then assessed – again on a scale of 1 to 10 – and a measure of confidence allocated to each value, once more on a scale of 1 to 4. A single score can be calculated for a particular course of action in a similar way to that described earlier.

The connectance model is rather similar, expressing a notion of how much a change in one aspect of a firm's operations causes a change in another. It is mostly concerned with identifying the strength of causation between different elements, whereas the rating system appears to concentrate more on people's judgement about where a firm's priorities lie. The first stage in building a model of this type is to identify a number of what are simply called aspects – significant elements such as sales revenue, lead-times, customer service capability and so forth. The next stage is to define connections between the aspects on the basis that changes in one have definable effects on another. The connections are weighted according to how powerful an influence the cause has on the effect, and an indication is given of the direction in which the influence takes place – whether an increase in the causing aspect gives rise to an increase or decrease in the affected aspect. Lehmann gives an example in which the scale of lead-times and inventories influence the success of customer service and, thereby, a firm's sales revenue.

A second, more complex example may be found in a paper by Srinivasan and Millen.[3] The method they describe is based on a relatively sophisticated scoring system known as the analytic hierarchy process (or AHP). Although, for the most part, the AHP follows some fairly common-sense notions about decision making, some of the manipulations used in the model are rather involved; I shall pass over these complications. The technique is described in detail in an article by Zahedi,[4] and this is a useful source for those wanting to understand both the guts of the method and the types of problem to which it has been applied. The model is not a linear one, but it is compensatory and subjective, so it is rather similar to the more rudimentary models described earlier.

There are four stages to carrying out an analysis with the AHP. The first is to write down a decision hierarchy – a tree-like structure that breaks down the facets of a decision into successively more detailed components over a number of levels. At the leaves of the tree are placed the different courses of action available to the decision maker. An example described by Srinivasan and Millen has a decision

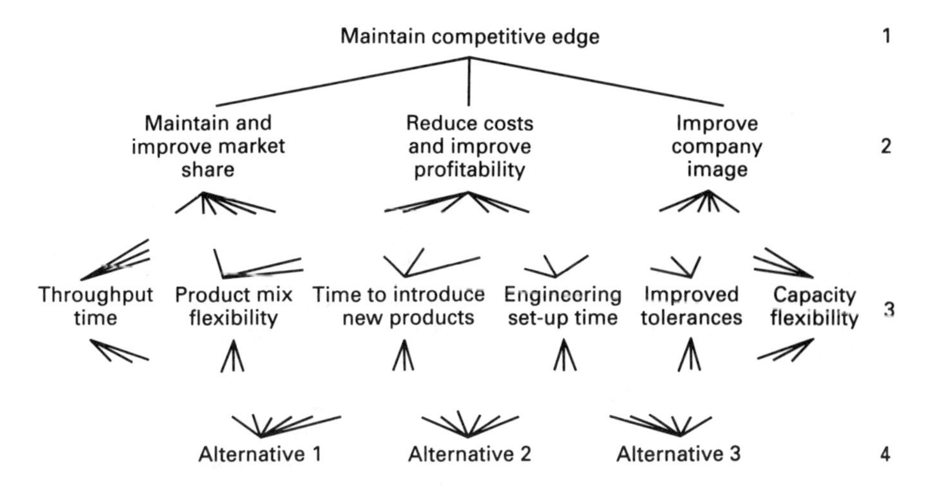

Figure 4.1 A decision attribute hierarchy

maker faced with the problem of choosing from three alternative developments in order to maintain a company's competitiveness. The sketch in Figure 4.1 summarizes their decision hierarchy.

This sketch indicates, for example, that choosing alternative 1 has an impact on various factors at level 3, such as throughput time. Throughput time in turn contributes towards several higher level properties, such as maintaining market share. Market share then has a certain impact on the ability to maintain a competitive position.

The second stage is to assess how much each factor at each level contributes to all the factors at the next level up, relative to the other factors at the same level. An analyst would, for instance, need to ask how much a reduction in throughput time helps the firm increase its market share, compared with increasing the flexibility in product mix, reducing the time to introduce new products and so forth. The distinctive element of the AHP is that it relies on pairwise comparisons. Rather than ranking the effect of throughput time on market share along a scale that goes from 1 to 10, the analyst ranks it *against* product mix flexibility, product introduction time and so on. The motivation for this is that it is easier and more robust to make such pairwise comparisons than it is to allocate numbers on an absolute scale. This is an attractive element of the AHP, but it has the effect of making the subsequent manipulation more complicated.

The result of this process is, for each level of the hierarchy, a series of matrices denoting the results of these pairwise comparisons. Every entry expresses the relative contribution to a higher-level issue of one factor compared with another. For example, if the ith factor is 5 times more important than the jth factor then the value in the (i, j)th position of the matrix will be 5. At every level, there is one such matrix for every one of the factors in the level above. The third stage of the process is to work out a list of ratings for the lower level factors against the higher level factor, in place of pairwise comparisons. This calculation is generally too involved to be done by hand, and the mechanism will not be described here.

The final stage is to aggregate the relative weights found in the preceding stage, with the intention of finding the relative impact of each of the alternative courses of action (at the bottom of the hierarchy) on the primary decision attribute (at the top). To do this, the lists of relative weights at every level are combined in matrices, and the matrices from each level multiplied together. The result is another matrix that can be broken down into a series of vectors, such that the kth vector is a list of the relative contributions of the factors of level k to the attribute at the top. Again this is a process that I wouldn't much like to do by hand.

There is in fact quite a bit more to the use of the AHP – the ability to test the degree of consistency in the weights attached at each level, for example. Srinivasan and Millen also combine the results of the AHP with a financial analysis. So this is plainly an involved technique, and it is one which you would be hard put to use without some sort of automated assistance. Whether the outputs are worth the complications depends on the tastes of the method's potential users: it is perhaps more for operational researchers than it is for industrial managers. And for all its sophistication it remains a scoring model, lacking a clear financial interpretation.

Drawbacks of scoring models

Scoring models are not especially good at making the evaluation process an objective one. They prompt people to state explicitly both the set of values they apply to decision making, and their perception of a development's effects, but there is no scale of value independent of the person who happens to be doing the analysis. Any decision maker must expect to have to compensate for the particular analyst who presents him with an evaluation, unless analyst and decision maker are one and the same person. If he is lucky, the analysis the decision maker is given might explicitly list the weights the analyst has attached to each of the several issues of interest. He can then either give them due credence, or substitute his own scale of priorities. (He has to do this at regular intervals, of course since the analyst's views will doubtless change as he learns more about technology, manufacturing and the world in general.) If all the analyst reveals is a final score his case will hardly be a compelling one.

It also means nothing to a firm's bankers or shareholders that System Z, for which the firm is trying to raise finance, has a net weighted score of 5. They are much more likely to be interested in a measure of worth expressed in monetary terms. The examples also suggest that these deficiencies are not necessarily compensated for by simplicity or clarity; scoring models can quite easily become complex and somewhat opaque.

This does *not* mean that a scoring system has no role to play in successful administration. It may well help people to put some structure to their understanding of costs and benefits, and it doubtless encourages them to make previously unwritten assumptions more explicit. Doing so probably helps build a degree of consensus among those to whom it falls to implement a new development. Scoring systems may also help bring into the open those unsupportable assumptions that time or changing circumstances have made invalid. So they are useful tools for making priorities explicit, and for helping a firm's technologists and managers think through where their sources of competitiveness lie. But in the final analysis they are not as objective or as consistent as we would like appraisal methods to be. They do not summarize the opportunity costs of taking certain courses of action; they do not use a common unit of measurement; they do not reflect a common scale of value. This makes scoring models an unsatisfactory approach to appraisal, whatever their appeal as a component of other management processes.

4.3 Rules of thumb

How can long experience of the financial properties that successful developments exhibit be incorporated in the evaluation of difficult and novel technologies? How should accounting information be used in the justification of new investments? How can it be established whether a new process technology will increase or decrease the measure by which the firm's managers are personally assessed? How can the appraisal be reconciled with the indicators of success that appeal to the people who control and monitor the company's funds? Some would say that the answer to each of these questions is to use one or other of the widely-known rules of thumb often associated with investment appraisal.

An example of such a rule might be that only those investments that promise a payback period of less than three years are to be adopted. Another might stipulate a return on investment of greater than 25%. These rules are based on simple yardsticks, such as payback period and return on investment, which are felt in some way to express the potential value of a proposed development. The rules are rather arbitrary, and largely dependent on local circumstances, and it is the yardsticks that will be of most interest here. Everybody understands such yardsticks, although not everybody may be quite sure about the assumptions on which they rest, or about the limits beyond which their expressive power shouldn't be pushed.

In calling payback period and return on investment thresholds *rules of thumb* I am following the terminology sometimes used to distinguish them from decision making criteria based on finance principles. You may feel that the use of a measure like the return on investment is sufficiently fundamental that it deserves to be categorized better than as a rule of thumb. But the term is not meant to be a pejorative one. It is meant simply to denote that such measures are relatively unsophisticated, and that they rely on experience rather than science for their cogency. The fact that they do not originate in a body of theory counts against them but it doesn't make them useless. The fact that experience may serve as a poor guide for the introduction of a new technology inclines us to be wary of them, but not to discard them altogether. As the opening paragraph suggested, it would be helpful to use evaluation measures that can make use of the information collected by accounting systems, and to have measures that are consistent with the bases by which managers are assessed by their superiors. In other words, there are practical issues as well as fundamental principles behind the process of appraisal, and rules of thumb have arisen in response to these.

A paper by Baumol and Quandt has suggested that rules of thumb always have a number of common characteristics:[5]

- the quantities they use are measurable in an objective way (this contrasts with the highly subjective ratings made in scoring models);
- the decision criteria are communicable, again in an objective way, and they don't depend on the judgements of individuals (this contrasts with the weights assigned to properties in scoring models);
- for every possible combination of relevant variables there is a definite decision and this decision is always the same one for that combination;
- the calculations are simple, inexpensive, suited to frequent application and suited to frequent spot-checking by managers.

Just two examples of the yardsticks used in rules of thumb are discussed here (the payback period and return on investment), but it is possible to think of many more than this. There is of course no finite collection of rules of thumb at all: new ones satisfying Baumol and Quandt's criteria can be invented at will.

Finding the payback period

The payback period of an investment is simply the time it takes to recoup the initial outlay – the time to completely restore expenses with revenues. It is usually calculated using cash flows, rather than quantities that have passed through the distorting processes of historical accounting.

Typically, when introducing a new technology, there are immediate cash outflows associated with buying hardware, software, peripheral machinery, networks and so on. The payments might, when the purchases are very large, be staged over an extended period, and there might in addition be one-off expenses for project management, training, consulting advice and so forth. But it is commonly the case that payments are completed before any returns start to be made on the investment. There will then be recurring expenses, perhaps leading to a steady cash outflow every year. The compensating cash inflows will, hopefully, continue to accumulate up to a point at which their size in aggregate has exceeded the total outflow: that point lies at the end of the payback period. The measure is such a well-known and simple one that it doesn't seem worthwhile giving an example.

Most companies incorporate the payback period in a single, specific rule. Often, the period will need to be less than either two or three years if a development is to be considered worth pursuing. In most companies this appears not, however, to be an inviolable rule. Opportunities with short payback periods are frequently allowed to lapse because a predetermined capital budget has been exceeded, because the project's sponsor is poorly thought of in the social circles that matter, or because there is some question about the credibility of the assumptions on which the calculations have been based. Sometimes, opportunities with an excessively distant payback are adopted: a plain indication that the yardstick doesn't capture every kind of information relevant to investment decisions.

It should be clear from the fact that the payback period is expressed as a time that it is not a measure of value – for few people would measure value in units of years. A period of time *might* ultimately be capable of translation into an expression of value, but that would involve further steps and quite a few new assumptions. Although the output is in time units, the inputs to the payback calculation are nonetheless financial quantities, and as such they are more easily tested than ratings on a scoring model.

The payback period clearly doesn't provide a comprehensive assessment of a proposal. A project that involved an outflow of £2000, and inflows of £1000 every year thereafter (for maybe five years) has the same payback period as one with an outflow of £2M and inflows of £1M. The second is of course far more valuable. It would be a very idiosyncratic firm that paid as much attention to the £1000 project. If payback were the beginning and the end of the matter, the alternatives would be apparent equals, since each breaks even in the same time.

The payback period is perhaps best categorized as being an indicator of liquidity (rather than one of profitability) since it measures the length of the period over which a firm is exposed to the risks and costs of funding a development. Until the period has come to an end, funds have to be found from elsewhere to support the project – whether they are taken from cash flows generated within the company or whether they come from outside. But it is hardly an informative measure of liquidity, since it would be much better to know the entire pattern of cash inflows and outflows. It is more a quick test than a definitive guide.

Drawbacks of the payback period

The main difficulty with payback periods is that they summarize only a part of the information that is potentially available about the expected pattern of cash flows: they stop short when the project has notionally broken even. This suggests that alternative courses of action will not be ranked satisfactorily, for there is very little reason to suppose that the most fruitful developments are inevitably going to pay back their initial investment the most quickly. Intuitively, you might even wonder whether the best developments aren't those that often prove difficult to get going.

An important question is whether the use of a payback yardstick might discriminate for or against advanced manufacturing technologies. It is probably fair to say that such technologies are at a marked disadvantage because of their novelty. Novelty suggests that managers have to spend time understanding new ideas, and either training or hiring new technologists. They have to devote time to overcoming people's inbuilt resistance to change. They have to cope with the loss of time associated with making mistakes that follow from their lack of knowledge about how the technologies are best installed and applied.

Whether the penalties are paid in terms of lost time or lost cash flows will depend on how the development is managed. A firm could trade profitability for liquidity, to some degree, by making extensive use of consultants. At the expense of their fees, consultants might help the firm to avoid time-wasting pitfalls. Either way, the payback period suffers – whether through delays in waiting for the benefits to take effect, or through increased development costs. In many cases there will be a justifiable scepticism that such developments will ever make a return, but there will be other instances in which work with a long gestation period proves highly worthwhile. It is not necessary to show that the use of payback periods always leads to the wrong decision in order to make a case for finding something better. It is only necessary to show that it *could* provide the wrong basis for a decision.

An additional difficulty with the payback period is that it doesn't incorporate any measure of uncertainty. There is no basis, for instance, on which to modify the output of the evaluation process to reflect a feeling that cash flow predictions are especially sensitive to future circumstances beyond a firm's control. This is a significant consideration because risk plays an important role in most people's perception of what is worthwhile undertaking.

Set against this is the interesting observation[6] that the payback period is less susceptible than other measures to biased forecasts. The reason for this is that any excessive optimism in the cash flow forecasts is likely to be concentrated in the time beyond the payback period – since the farther things lie in the future, the less easily they can be contested. The payback period rule of course ignores everything that happens after the end of the payback period, so this bias tends not to affect the calculations.

On balance, however, the payback period provides a very limited evaluation of the opportunities open to a firm, and it is likely to throw an unjustifiably harsh light on new technologies. There are various explanations for the persistent popularity of the payback period: a paper by Statman[7] has some interesting ideas. Most people recognize its limitations, but this has not in many firms loosened its hold on the decision-making process.

Finding the accounting rate of return

The accounting rate of return (ARR) more often goes by the name of return on investment, but it is quite common to use the term ARR when conducting an investment appraisal. ARR is in fact the more expressive term since the most important characteristic is its basis in accounting measures.

As with the payback period, it is easy to understand and generally straightforward to apply. The first step is to find the average level of profits over the lifetime of an investment according to normal reporting conventions. This means, for example, that a sum has to be deducted from the earnings in each period for depreciation – to reflect the fact that fixed assets gradually lose their earning power. It is, of course, a fictitious expense that bears no relation to cash flows. It is calculated on some predetermined basis, perhaps by taking a fixed percentage of the initial investment, or by taking a percentage of the balance that remains.

The average profit level is generally taken to be a mean. In other words, the profits of each year are added up, and then divided by the number of years in the life of the investment. Suppose, for example, that before considering depreciation a project's annual earnings are £2M. Suppose also that there is an initial investment of £5m, and that the lifetime of the development is five years. After five years there is no earning power left to speak of. Then using straight-line depreciation the average annual profit is £2M – (£5M/5), or £1M.

To find the ARR, the average profit is divided by the average capital investment. Since the net value of the investment decreases along a straight line, the average is the mid-way point between £5M (the capital asset value at the start) and zero (the written-down value at the end) – that is, £2.5M. Hence the accounting rate of return is £1M/£2.5M or 40%.

The acceptance criterion associated with the ARR is often that among a number of choices the one with the highest ARR is adopted. An alternative is that the ARR of chosen projects has to exceed a predetermined hurdle rate, say 25%. In practice it is sometimes the decision to accept or reject a project that is predetermined, not the hurdle rate.

Pros and cons of the accounting rate of return

The main advantages of the ARR are perhaps its simplicity, and its consistency with the way in which managers of investment centres are often assessed. The use of a return on investment yardstick both for evaluation prior to investment, and for reporting after the event, makes some sense. It means that people can get a feel for the direct effect of a new development on their own bottom lines.

Among the drawbacks of using the ARR are its basis in accounting conventions, the fact that it ignores cash flow timings, and its inability to incorporate a measure of uncertainty in the forecasts on which it is based. The problem with the accounting way of thinking – at least that of *financial* accounting – is that costs and revenues are re-allocated to the periods in which an associated activity takes place: that they are matched, year by year. For example, a machine with a five-year life might be expected to wear out one-fifth of its earning power in each of the five years following its purchase. A charge of one-fifth of the machine's purchase value is, in this case, made against the firm's profits in those years. But, from the point of view of a manager making the decision about whether to invest in the machine, it is much more significant that all the cash needed to buy it leaves the company before the end of the first year. He has to arrange for this cash to be available, and he will pay for his indebtedness, well before any associated revenues start to enter the firm. Even if the cash is generated internally there is an opportunity cost associated with applying it – the benefit foregone, say, by not putting it in the bank to earn interest. To the decision maker, a notionally fair allocation of the loss in value of that machine is neither here nor there; it is cash flows and timings that are important.

In addition, the ARR is, as its name suggests, a proportional measure. If a firm has the option, say, of investing in a development that promises a 40% ARR, or one that promises 20%, the temptation might always be to choose the first. It is quite possible, however, that the first is expected to earn absolute profits of £2000, and the second profits of £2M. Invariably selecting the opportunities with the highest ARR would sometimes be an absurd rule to apply – the ARR simply doesn't convey enough information.

Finally, it is worth considering whether there is anything about the use of the ARR that discriminates for or against advanced technologies. Having to negotiate rather arbitrary hurdle rates is something of a lottery, and it could be argued that newer technologies, with their attendant costs of introduction, are likely to show a relatively low rate of return and perform poorly by ARR. There is nothing intrinsically wrong with this – the whole point of the evaluation is to discriminate against investments that promise a low return. But is there an associated, *unjustifiable* discrimination? The main reason why there could well be is that it is tempting to omit the accounting effects of radically new systems because they are too difficult to quantify.

This fate will be suffered at the hands of any quantitative method, but it will be especially severe with one founded on accounting principles. Not only do accounting yardsticks establish particular ways of calculating certain quantities, but they also place an emphasis on using historical information. This is supposed to increase the objectivity of figures presented to outsiders, on the basis that whatever is documented and historical is less open to dispute than anything that is undocumented and untested. In reality, the objectivity of an appraisal is compromised as much by errors of omission as those of commission. So although it might make sense to discount partly the potential benefits of a new technology because they are uncertain, it wouldn't make sense to discount them completely because they are of a type that hasn't been measured in retrospect.

Some closing remarks on rules of thumb

One way of attempting to improve the applicability of rules of thumb is to combine them. If the payback period doesn't measure value, and if it fails to incorporate sufficient information, then perhaps it should be used jointly with the accounting rate of return.

The first drawback of doing this is that it spoils the main benefit of rules of thumb – their simplicity and the speed with which they can be worked out. Another problem is that the outputs of two yardsticks in combination are rarely comparable: the output of a payback calculation is a time period, while that from an ARR calculation is a percentage return. It isn't possible, without some additional apparatus, to say how many years of shorter payback are worth one extra per cent of ARR.

Nor is it satisfactory to say that rules of thumb should simply be used to augment the more firmly-founded financial measures described in later chapters. The payback period rule discriminates against developments with a long gestation period, and it may rule out projects which, when measured against the financial yardstick, turn out to be highly significant. In other words, it would apply a filter to the evaluation process that would sometimes remove proposals which ought to be retained.

The simplicity and cheapness of rules of thumb considered individually suggests that they might still be better than more complex approaches, even if their benefits are somewhat smaller. Unfortunately the problem is not just that there is a *random* uncertainty about how far decisions based on rules of thumb would differ from those made under more comprehensive analyses. They also introduce a significant *bias* to the decision-making process: irrespective of the values held by the people that apply them, these rules embody principles that are bound to favour small, short-run developments that mimic existing systems.

Another argument occasionally raised in favour of rules of thumb is that to accuse companies of failing to understand the full sophistication of financial principles is not the same thing as accusing them of failing to make money successfully. There are many instances in this life of people and institutions excelling at things they cannot properly explain. It could be that the motivating power of simply-expressed targets – like maximizing the return on investment – is more important than having a complete understanding of all the associated complexities.

There is even a possibility that with too great a degree of expertise a decision maker might become so overwhelmed by the need for rigour as to be unable to make bold and effective decisions. We often hear of successful (but disarmingly modest) people who say, in hindsight, that if they had understood when they set out what a difficult undertaking they faced they would not have started in the first place.

What don't register so powerfully, perhaps, are the instances where a superficial understanding of the way the world works has led to failure. Nor are these occasions likely to be so well recorded for posterity. An article by Kim[8] from a few years ago reported the results of a survey that attempted to determine whether there was any correlation between the earnings of a number of American companies and their appraisal practices. He found in fact that there was a positive relationship between the degree of sophistication of a firm's capital budgeting processes and its performance. (His paper provides definitions for these quantities.) Correlation does not imply causation: the findings in no way prove that sophisticated appraisal is good for a firm. But they are an indication that doing things the least expensive way is not a satisfactory end in itself. Simplicity without expressive power, consistency and relevance is worthless.

There is no reason to suppose that the process of evaluating difficult and complex technologies *should* be straightforward. Most other aspects of their management – the organizational issues, the integration with other technologies, the maintenance and operating problems – are really quite hard. And because rules of thumb arise out of experience they are only likely to be useful while that experience continues to be relevant. Since we are now contemplating substantial and wide-ranging changes in both markets and technology, this makes it highly desirable to find systems that measure financial success more directly.

4.4 Summary

There are several informal approaches that one might take to carrying out an appraisal of a new system. Since they are based on practical ideas, rather than a rarified set of principles, the tests of a good approach are themselves very practical. We should look for:

- a way of setting benefits against costs by reducing each to a common scale;
- a way of recording the uncertainty with which we make predictions, and of incorporating it in such a way that it influences the decision rule based on the chosen system of appraisal;
- an approach that provides results which are stable from one time to another, and consistent from one person or group to another.

One set of informal approaches to making investment decisions falls under the heading of scoring models. These are mostly based on a procedure in which the analyst:

- lists the properties of a new development that are relevant to the investment decision;

- assigns each of these properties a weight to reflect its relative importance;
- rates the alternative courses of action being considered against each property;
- forms a total score by adding the products of individual ratings and weights; and
- selects the alternative with the highest score for adoption.

A second type of decision-making system is the rule of thumb. Rules of thumb, like scoring models, are generally cheap and quick to apply, but they are distinguished by:

- their use of data that can be measured objectively;
- their use of decision criteria that do not depend on one person's judgements.

These rules are often based on testing whether a proposed development crosses a given threshold expressed either in terms of a financial measure (such as the payback period) or an accounting yardstick (such as the return on investment).

Informal approaches have a number of drawbacks. None of the examples described in the body of this chapter, for instance, satisfied all the tests specified at the beginning. Although we can't examine in a quantitative manner whether these drawbacks more than offset the advantage of cheapness, it seems highly likely that they in fact do so. Among the drawbacks is a characteristic bias against new technologies – a bias that does *not* reflect an objective measurement of the difficulties of applying new technology. It is built into the structure of these informal approaches.

Notes and references

1 CAM-I, *Management Accounting in Advanced Manufacturing Environments – A Survey*, for CAM-I by Coopers & Lybrand, Ernst & Whinney, Peat Marwick McLintock, January (1988)

2 Lehmann, S. How to change intangibles to tangibles in CIM-analysis. Paper to the CIM-Europe Conference. Madrid 18–20/5/88

3 Srinivasan, V. Millen, R. A. Evaluating flexible manufacturing systems as a strategic investment. In Stecke, K. E. and Suri, R. (eds.) *Proc. 2nd ORSA/TIMS Conference on Flexible Manufacturing Systems*, Elsevier pp. 84–93 (1986)

4 Zahedi, F. The analytic hierarchy process – a survey of the method and its applications. *Interfaces*, **16**(4), 96–108 (1986)

5 Baumol, W. J. and Quandt, R. E. Rules of thumb and optimally imperfect decisions. *American Economic Review*, **54**(2), 23–46 (1964)

6 Franks, J. R. and Broyles, J. E. *Modern Managerial Finance*, John Wiley & Sons, p. 36 (1984)

7 Statman, M. The persistence of the payback method: a principal-agent perspective. *Engineering Economist*, **27**(2), 95–100 (1982)

8 Kim, S. H. An empirical study on the relationship between capital budgeting practices and earnings performance. *Engineering Economist*, **27**(3), 185–95 (1983)

5 The present value yardstick

... several excuses are always less convincing than one.

Aldous Huxley *Point Counter Point*

The four sections of this chapter look at a single, comprehensive yardstick based on finance principles. The first section describes the most relevant of these principles in a qualitative fashion, and the purpose here is to convey a line of reasoning rather than a detailed procedure. It is this that has a broad, unchanging applicability: and when the practical procedures have outlived their usefulness as a result of changing circumstances one can return to this reasoning to work out how the procedures should be modified. The second section describes how calculations are carried out in practice. Because of the costs and difficulties of collecting information about future developments, and of the need to make the basis of the results easy to grasp, a number of simplifications have to be made at this stage. The final sections consider the main characteristics of computer-based systems in terms of their effects on this financial yardstick.

Some of the ideas in these sections seem questionable at first. There is a clear emphasis on using the appraisal to say whether a firm is doing the best for its owners, and there is no explicit recognition of the interests of other people who may be concerned about the firm's performance. Employees, for instance, would probably not like to feel that their interests were being ignored. It would be tempting to say that financial approaches cannot be wholly valid because they only capture quantities that are valued by shareholders; they fail to pay any attention to the desires of workers for a livelihood and rewarding employment. In some detailed respects this may be true. As will become evident, the way in which shareholders look upon risk is likely to differ from the way in which managers and workers do so. But the fundamental purpose of appraisal – to find the way of using resources that can be expected to offer the biggest premium of benefit over cost – is *not* concerned with just one group of people's well-being. (How the rewards of making money are subsequently distributed is another matter.) To say that the arguments are cast entirely in favour of shareholders would also ignore the extent to which rewarding employment depends on a firm's financial success, and the extent to which a firm seeking financial success must provide rewarding employment.

There is nothing especially original in the account given in the earlier sections. You can find much fuller and more closely argued descriptions in general texts on finance theory,[1] or on financial management.[2] But there is enough here to serve as a foundation for practical applications, and for the discussion about computer-based systems in the later sections.

5.1 Underlying ideas

The purpose of financial appraisal

A good starting point is to review the basic financial process that takes place in an industrial firm. The essential form of this process is that a firm acquires money in highly competitive financial markets, and uses it to exploit opportunities in product markets in which there are few or distinctly inferior rivals. Because financial markets are so competitive, a firm can expect to pay a fair price for its funds – a price close to that of any other investment with similar relevant characteristics. Because product markets are less competitive, on the whole, a firm can exploit an initial advantage of some sort. It might be a profound understanding of consumers' desires, or a special ability to organize engineering and manufacturing operations. Such advantages are usually temporary, in the absence of sustained actions to develop them, because in time they tend to be eroded by competition.

Looked at in these terms, industrial investment is a kind of arbitrage activity:[3] people who think capital is more effective in the form of a manufacturing establishment than in the form of cash will acquire cash and exchange it for the machinery and know-how of manufacturing. They will do so in the expectation that the returns to manufacturing exceed the charges they pay for finance. The process continues while manufacturing remains more lucrative than the other activities to which the same capital resources could be applied.

The point of conducting a financial appraisal is to predict the success of this process of converting capital into the apparatus of manufacturing. The decision maker wants an estimate of the value gained by applying financial resources to a new manufacturing development, and he wants a summary of the value that is foregone by denying those resources to alternative applications. The difference between these two values is the net worth of the development.

It would be very hard, in practice, to find all the alternative ways in which capital could be applied. However, capital is traded in free-ish markets, and in these markets the forces of competition will mean that the price of capital reflects the desirability of the different opportunities for investing it. It is therefore unnecessary to know in detail about all the opportunities that are foregone when capital is applied to one particular development. They are summarized in the price that the market sets.

A company cannot reduce the opportunity cost of its funds by generating them internally – by taking them out of its profits instead of issuing shares or selling debt. The company can place its own funds in the financial markets instead of applying them to internal developments, and it would only make sense to proceed with internal developments if they promised a higher return. So wherever the funds come from, the opportunity cost of establishing a project is determined by the market price of capital.

This price will, of course, vary according to the nature of the scheme in which capital is being invested, and the initial task in an appraisal is to determine the value it takes for a specific advanced manufacturing development.

The opportunity cost of capital

The first thing to note is that people, for the most part, have a preference for things sooner rather than later. There is an intrinsic value in the opportunity of acquiring a commodity now rather than at some time in the future. There are various reasons for this:[4] the fact that the future is not so vivid as the present, for example, and the risks and contingencies of life that mean we may not see much of the future at all. The result is that people who temporarily provide others with resources (including money) charge them interest. They do so to an extent that reflects the loss in value the lender feels by postponing the use of a resource to some point in the future, and the increase in value the borrower feels by bringing forward the use of that resource to the present. Since it is too difficult to work out this time value from basic psychological principles, it is normal to look at capital markets to see what prices they establish in the process of reaching an equilibrium between the desires of sellers and buyers. In general, the longer the time for which resources are provided, the greater these prices are.

The second determinant of our cost of capital stems from the fact that most people are averse to risk. Suppose someone had the choice either of receiving £100 with absolute certainty, or of receiving the outcome of a lottery that promised – with equal likelihood – a payout of £70, £100 or £130. The *expected* value in the two cases is the same: the first is obviously £100, the second (1/3 × £70 + 1/3 × £100 + 1/3 × £130) is also £100. But choosing the second option runs the risk of receiving only £70. This risk arises because there is insufficient information to narrow down the outcome to a single value, and it is independent of the expected, or mean, value. The fact that someone is risk averse means that he will choose the certain return in preference to the lottery.

Similarly, a person providing funds to others will commonly prefer to direct them to applications in which there is the least breadth of uncertainty in the returns that he will make. So risk implies a loss of value. This loss interacts with time value, or the loss due to postponement, in the sense that in many instances the longer the returns are postponed the greater the risk that they will deviate from the expected value. The two effects are, however, distinct and they have different sources.

An obvious indicator of risk is the spread of probabilities that we attach to future events. In our simple lottery it was the distribution of probabilities among three distinct outcomes that made the lottery less valuable than a certain result (where there was no spread at all). We would probably have judged the risk to be still greater had the distribution been one in which there was a 50% probability of receiving either £50 or £150 (even though the expected value is still £100). A measure of this spread is the variance, and we might suppose that the greater the variance in the anticipated returns to an investment the lower the value, all other things being equal, of the asset that generates these returns. This will not necessarily be easy to calculate in practice, but since this discussion is still framed in abstract terms it might do for the moment.

Unfortunately the analysis is complicated by *portfolio* effects. These are the results of holding more than one asset at once – that is, of holding a portfolio. If the returns on the assets in a portfolio respond in a different fashion to the influences that affect them, then the returns on the portfolio in total will be less sensitive to these influences than those of the assets considered individually. In other words, the holder of a portfolio can, by diversification, reduce the risk he faces. It is natural, therefore, for the providers of capital to build portfolios of assets selected as far as possible to exhibit fluctuations that offset one another. It is most unlikely that a portfolio can be built in which the assets' returns are perfectly negatively correlated because some factors, especially the general economic climate, have the same type of effect on every business. But it is possible to diversify away a certain amount of risk without suffering a reduction in returns.

The upshot is that it is not the absolute level of uncertainty about its returns that makes an asset less valuable; it is, instead, the extent of the uncertainty that cannot be removed by holding a diversified portfolio of which that asset is but one member. We can therefore expect markets in capital assets only to set a premium that rewards a provider for the non-diversifiable elements of risk in an asset's returns. The provider can, at very little cost, remove the influence of diversifiable risk.

As a result, an industrial company calculating the opportunity cost of capital should not attempt to incorporate risk in total. The cost of capital arises only from the non-diversifiable part. This is a very important fact. It means that some developments whose effects are very uncertain may not be penalized for this uncertainty if the non-diversifiable element of the risk is slight.

If we want to find the risk premium for a given development we need two pieces of information. The first is the price that capital markets set for each degree of non-diversifiable risk. The second is the degree of non-diversifiable risk associated with the development in question. Exactly how we might obtain this information is the subject of the next two sub-sections.

The market's valuation

The most common way of looking at how capital markets set prices is to use the capital asset pricing model. Its essential features for our purposes are illustrated in Figure 5.1. The horizontal axis describes a scale of non-diversifiable risk and the vertical axis the expected percentage return on an asset. The straight line shows the relationship between risk and return established by the market under equilibrium: if we can say what we think the risk of an investment is, we can read off the expected return. This expected return represents the opportunity cost of capital as a percentage of the capital's value over a single period.

When the risk is zero, the line crosses the vertical axis at the risk-free rate of interest, or R_f. This represents the cost attributable purely to postponement. The way in which the horizontal axis is scaled is of no concern here, but the measure of risk is commonly known as β from the statistical process that lies behind its definition. When we are considering the operations of a firm as a whole we can simply look up the value of β in published tables, provided that the firm's shares are listed.

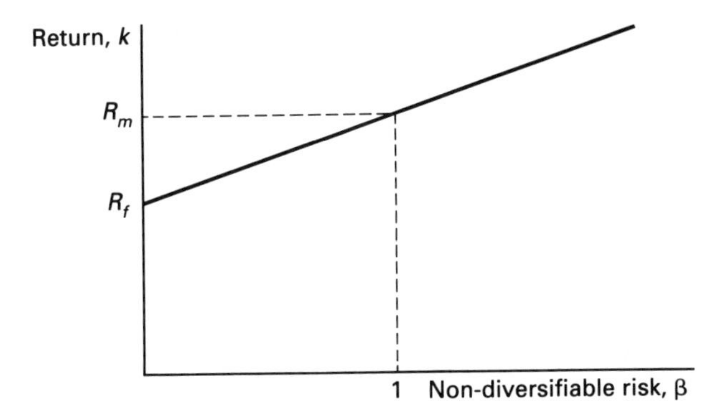

Figure 5.1 Rudiments of the capital asset pricing model

To define the straight line fully, we use the point at which β is one. The corresponding value on the vertical axis is known as the market return, or R_m. One can think of R_m as being a kind of average return in the market, and of unity as being an average level of risk. The equation of the straight line, and therefore the expected return on an asset, is thus

$$k = R_f + \beta(R_m - R_f)$$

(Strictly speaking, R_m is a random variable and we should speak of its expected value, or $E(R_m)$.)

For practical purposes, the easiest way of using this formula is to assume that $(R_m - R_f)$ takes a constant value of 6%. This, arguably, is its historical average.[5] R_f can be taken as the prevailing rate of interest on government securities: it is again arguable whether these should be long-term bonds or short-term bills. Our equation is now simply

$$k = R_f + 0.06\ \beta$$

If we can find a value for β we therefore have a quick way of finding the expected return on a given capital asset, and thus the opportunity cost of employing it. For example, if the risk-free rate were 10%, and we were looking at an asset with a β of 1.5, the expected return on that asset would be 10% + 1.5 × 6%, or 19%.

This model, and the assumptions that have been made here, are by no means watertight from a theoretical standpoint. But we simply want to find a representative value for the opportunity cost k. We are much less worried about the way in which the market forms its prices, and about effects that will have only marginal influences on our calculations. In any case, the impact of any refinement to this model will probably be dwarfed by uncertainties in the raw data that will have to be fed into it.

Risk and the single project

We now have a way of finding the cost of capital once we know the value of β. Since our interest lies in the worth of a specific development, it is the β that is applicable to this development that we need to calculate. The easiest way of doing this is to use the development's characteristics to modify the firm's β. The firm's β can be looked up in a book if the firm has a listing, or, if not, it can be worked out by comparing it with listed firms having a similar operational nature. (An example of this procedure will be given in the next section.)

Now the element of risk that cannot be diversified away is associated with general economic conditions whose effects very few people and institutions can escape. So we would expect to have to multiply the firm's β by the project's particular, relative sensitivity to these effects. In other words, the project β (or β_p) will be related to the company β (or β_c) by a relationship like

$$\beta_p = \beta_c \times \text{(project's earnings sensitivity/company's earnings sensitivity)}$$

If the company is of average risk (and has a β of one), and if the project's returns are twice as sensitive to economic fluctuations as those of the company in general, then β_p is 2.

In practice this sensitivity is hard to estimate directly, and it is better re-expressed in the form of two separate factors. Instead of asking directly how sensitive a project's returns are to economic changes, we ask how sensitive returns are to revenues, and then how sensitive revenues are to economic changes. We can speak of

$$\beta_p = \beta_c \times \left(\frac{\text{project's revenue-to-economy sensitivity} \times \text{project's return-to-revenue sensitivity}}{\text{company's revenue-to-economy sensitivity} \times \text{company's return-to-revenue sensitivity}}\right)$$

Or,

$$\beta_p = \beta_c \times \text{(relative revenue-to-economy sensitivity} \times \text{relative return-to-revenue sensitivity)}$$

A procedure along these lines has been suggested by Franks and Broyles and we can use their notation for the two risk factors:[6]

$$\beta_p = \beta_c \times f_1 f_2$$

The first of these factors, f_1, represents the volatility of sales revenues under changing economic conditions and it is concerned with the company's external, commercial position. It is measured relative to the same effect for the firm as a whole. The second factor, f_2, is the rate at which a project's returns change with sales revenues, and it is determined by the nature of a firm's internal operations. Again, this is expressed in a form relative to the sensitivity of the entire firm's earnings to its revenues. The manner in which this idea might be used is illustrated in the next section.

The first factor tends to be low (and certainly less than one) if a development is aimed at reducing fixed costs. By definition, fixed costs are independent of sales,

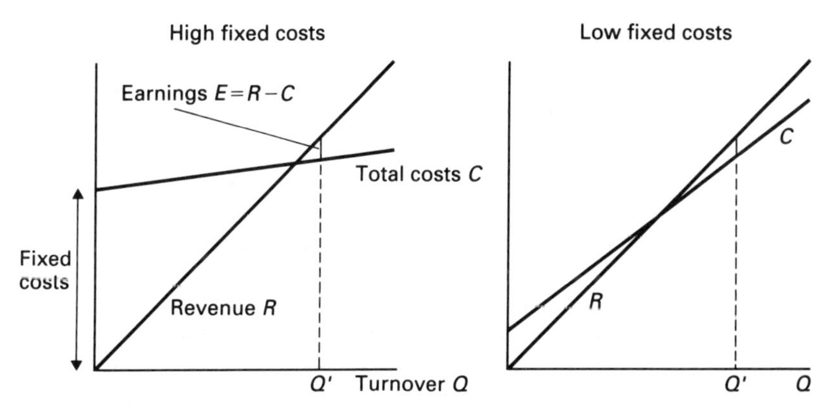

Figure 5.2 The effect of gearing on earnings

and if you can save £1M of some fixed expense you will save it, other things being equal, whatever the turnover the company generates. If, however, changes in the condition of the economy are expected to change the project revenues more than the company's revenues in total, then this relative sensitivity will have a value greater than one. And, as a result, the risk premium attached to the project will be higher than that of the company in general.

The second factor, the relative sensitivity of a project's earnings to its revenues, arises from having to meet fixed expenses. Figure 5.2 is intended to indicate why – if this is not apparent – fixed expenses increase the sensitivity of earnings. Suppose a company expects to have a turnover of Q units, and suppose it has the two options indicated in Figure 5.2 of how it may combine fixed and variable costs. The line R indicates revenues (simply a constant unit price multiplied by the turnover), and the line C indicates the total cost, fixed plus variable, for a given size of Q. At the expected turnover, Q', the company's earnings (the difference between revenue and total costs) are the same in the two cases. However, if Q' varies then earnings vary according to the gap between lines R and C. It may be seen from the sketch that where fixed costs are proportionally higher, there is a scissors-effect that means that a given change in turnover produces a bigger change in earnings.

The scale of fixed operating costs in relation to total operating costs is called operational gearing. When times are good – that is when turnover is increasing – a high level of operational gearing tends to magnify the rise in profits: when times are bad, it tends to magnify the fall. In other words, the sensitivity of earnings to turnover is increased by gearing.

The two factors involved in non-diversifiable risk, revenue sensitivity and operational gearing, may seem a little detached from the real uncertainties of new technology. But it is these two factors that influence the cost of capital; and it is therefore they that are relevant to the appraisal. The manner in which broader notions of risk are significant in the general process of managing new technology is discussed in Chapter 8.

Discounting the future

How then is the opportunity cost of capital incorporated in an appraisal? The most usual answer (although not in principle the most satisfactory[7]) is that it is used to discount future cash flows. The result, after discounting, is a *present value* – which, as its title suggests, is immediately comparable with cash flows occurring in the present. The extent of this discounting reflects the degree to which time preference and non-diversifiable risk make future cash flows less worthwhile than those in the present.

How do we know what a future cash flow is worth in the present, given the cost of capital? The easiest way to answer this is to use our knowledge of a similar problem – finding the worth of a current cash flow in the future. This is simply done by accumulating interest. If we received a single cash flow of £1M now, we could invest it in assets that, say, exhibited a level of risk such that the expected return was 15%. Then in one year's time this cash flow would be worth £1.15M, and in two year's time £1M $\times$ 1.15^2, or £1.3225M. We might equally suppose that a cash flow of £1.15M in one year's time, or one of £1.3225M in two years' time, would be worth £1M now – provided that the asset generating these cash flows exhibited a level of risk that matched the 15% cost of capital.

In other words, when we ask what is the present value of a future cash flow of C in n years time, we are asking what is the cash flow right now which, when it is inflated by compound interest, would yield exactly C within n years. The present value is therefore

$$V = C/(1 + k)^n$$

where k is the cost of capital. As we have already seen, k is related to the risk of the development from which these cash flows arise by the relation

$$k = R_f + 0.06\ \beta_p$$

And, again as we have seen earlier, β_p is related to a firm's β by two sensitivities:

$$\beta_p = f_1 f_2 \beta_c$$

The *net* present value (NPV) of a particular course of action is the quantity that remains after all relevant cash flows have been discounted by an appropriate amount and then accumulated year by year. For a development having a cost of capital of 15% which called for an initial investment of £1M, and then provided returns of £0.5M for each of the four subsequent years, the NPV would be

$$\begin{aligned} &-£1\text{M} + £0.5\text{M}/(1 + 0.15) + £0.5\text{M}/(1 + 0.15)^2 + £0.5\text{M}/(1 + 0.15)^3 \\ &\quad + £0.5\text{M}/(1 + 0.15)^4 \\ &= £0.43\text{M} \end{aligned}$$

The minus sign is used to distinguish a cash *out*flow. In symbols this can simply be expressed as

$$V = \sum_i C_i/(1 + k)^i$$

where C_i is the net cash flow in the ith year and k is the discount rate. (The C_i ought to be marked to indicate that we are using expected values of what are in fact random variables. In practice, the cash flows are quantified by simply making best estimates, not by taking averages of measurable distributions, so there is little point in adding such markings.)

It is the net present value that will serve as the fundamental yardstick by which the worth of a development is to be expressed. This yardstick is commonly used as the basis for a decision-making criterion that calls for the maximization of economic value. This stipulates that all, and only, those opportunities that promise a positive NPV are to be pursued. When there are mutually exclusive alternatives, the alternative with the greatest NPV should be chosen. This rule is a very strong one, for it implies that there is a perfect way of searching for investment opportunities. It also implies that there are no factors relevant to investment decisions that cannot be captured either in cash flows or in non-diversifiable risk. How this might be relaxed in practice is discussed in Chapter 7, and for the time being it is the present value yardstick, rather than the value maximization rule, that is the important issue.

Cash flows

The monetary values to which the discount rate is applied must be quantities that reflect exactly the anticipated effects of pursuing a particular development. There are, as a result, a couple of simple tests that these quantities should satisfy.

First, they should be relevant: they should hinge on the outcome of a decision. Effects which are unconnected with a particular development, or which are connected with it by anything other than a causal relationship, should be ignored. You might want to attribute a proportion of a factory's undivided overhead to a new production cell, for example. But if the decision about whether or not to proceed with the cell does not affect parts of that overhead, such as the cost of re-decorating the plant director's toilet, it should not be burdened with them. Such a cost is, of course, entirely unconnected with the decision about the new cell.

The second requirement is that these quantities should be cash flows. They should not have suffered any process in which the magnitude or timing of cash entering and leaving the firm has been adjusted. The accounting principles of matching and accrual are therefore inappropriate here, whatever their merits for historical reporting. The cash costs associated with making work-in-process are incurred around about the time of manufacturing, not when the finished goods into which they are converted are finally sold. The cash costs of capital plant flow when the supplier is paid, not when depreciation is charged. The reason it is important to observe this principle is of course that the timing of cash flows is highly significant: a cash out flow of £1M in Year 1 has a different economic value from a series of outflows of £200000 in Years 1 to 5.

When we look at cash flows in more detail in a later section of this chapter, some of the difficulties associated with the application of these tests will become evident.

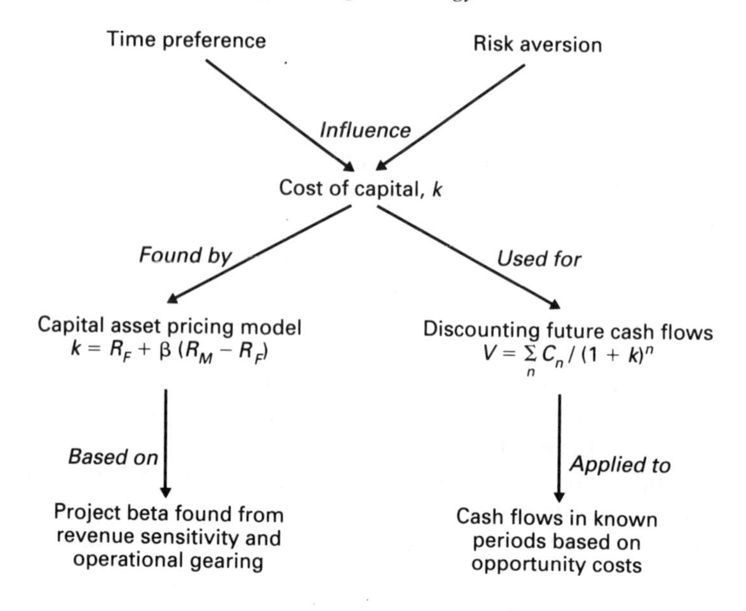

Figure 5.3 Summary of the present value background

Figure 5.3 summarizes the main themes of this first section. It is not, of course, necessary to keep all of these at the front of your mind when assessing real developments. But it probably helps to turn to the underlying explanations every now and then in order to check that the yardstick continues to be applied in a sensible way.

It should be evident that the NPV construction is not an arbitrary yardstick. It has not been developed on the basis of simple intuitions or common sense alone. It is not just a tool which you may happen to choose in preference to any number of other tools depending on your state of mind and the nature of the job to which it is being applied.

A development's net present value is a fundamental statement of its worth. It captures basic, verifiable ideas about how capital resources are valued. This doesn't necessarily make the calculation of present value an entirely practical procedure, but because it makes so comprehensive a statement about an investment's value it would be very appealing if it could be incorporated in the process of industrial decision making.

5.2 Calculations in practice

An example

Since we are concentrating here on the structure of the present value yardstick, technological issues will be ignored in this section. The workings will assume that we have a ready-made proposal for a new development, complete with cash-flow predictions and an understanding of the quantities that influence risk.

Suppose that we have a proposal for a system that calls for equipment purchases of £500000 over a period of nearly a year. Suppose also that we must spend a further £300000 on consulting, training and some external development work: we will need, perhaps, to have the system tailored to the particular conditions in our own firm. The firm will also have to spend £100000 on infrastructure – on offices, power supplies, air-conditioning plant and so forth. Finally, we have estimated a need for some 4000 hours of internal development and administrative work. Since this will be a service generated within the firm, we cannot assign it an objective price in the way that has been done for the purchased technology. We will in fact *assume* that the opportunity cost of this service – the value of its next best use – is equal to the recovery rate the firm uses in its reporting systems. This might be equal to the budgeted cost of running its computer department, say, divided by the number of hours of work it can trace to specific projects. Suppose this is £25 an hour: then the cash outflow associated with the internal development is £100000.

The use of recovery rates as a rough estimate of opportunity costs is our first explicit simplification. The second will be to attribute all the cash flows occurring in a particular year to the beginning of that year, which probably suits our margin of uncertainty about when things would happen in practice. By lumping together cash flows in this way, we can avoid the need to worry about discounting over fractions of a year. We therefore say that *at* Year Zero, there is a cash flow of minus (£500000 + £300000 + £100000 + £100000), or −£1M.

Over subsequent years, we shall assume that operating expenses are reasonably constant. In effect, we have given ourselves the first year both for installation and for learning – that is, for reaching a stage of some stability. Suppose in fact that there are constant recurring outflows of £100000 annually.

The benefits we should look for are *incremental* quantities, in the sense that they represent the difference between cash flows with the development in place and those without it. If, without undertaking the development, the firm expected its revenues to deteriorate by £10000 a year under the influence of strong competition, and if by undertaking the development it could expect to improve on the existing situation by £500000 a year, then the annual incremental benefit is £600000.

We also need to know for how long the project will last. In other words, we need to know the lifetime beyond which the firm will not experience any cash flows (revenues or expenses) caused by pursuing the project. There are various ways of determining project lifetimes, and the most satisfactory is to use economic arguments: to find the period that maximizes the projects's NPV. It is common, however, to take a short-cut and to use technological factors instead. Typically, these will be associated with obsolescence. A newer technology may be sufficiently attractive to make it worth ditching the existing one, and suppliers may become grudging about continuing to service systems that have become old hat. Although this is not as rational a basis as economic lifetimes on which to set limits to a project's longevity, technical and economic issues are strongly enough connected in most cases that they will yield similar results.

Even these arguments are usually clouded in practice. There are differing degrees of obsolescence and there are ways of modernizing a technology that saves replacing it. Some systems actually have benefits that outlast their physical presence – the experience and knowledge that people gain in their development and operation, perhaps. (How such an effect might be incorporated in an evaluation will have to wait until Chapter 6.) So what often happens is that firms arbitrarily fix *ex ante* lifetimes for their computer systems, at least for the purposes of appraisal. Four or five years seems to be a typical value. This is in fact our third assumption: that we can rely on the firm's predetermined lifetime of four years following a system's introduction (which itself takes a year). In each of the four years in which the project is operating, there is a cash flow £600000 – £100000, or £500000.

So, if we call the present moment the end of Year 0, we can draw the cash schedule (in £000s) thus:

Year 0	Year 1	Year 2	Year 3	Year 4
–1 000	500	500	500	500

The next stage is to find an appropriate discount rate, and, in order to do that, to find a value for the project's β. First, a value is needed for the company's β and here we shall assume that the company's shares are *un*listed, but that there is a competitor which in most material, operational respects is similar, and which does have a listing. We can therefore look up its current β in one of a few commercially available books, and we shall suppose that it has a value of 1.2. (An example of such a book is the publication of the LBS's Risk Measurement Service.[8]) There is, however, a complication in drawing a comparison between the risk associated with two companies: their levels of *financial* gearing. The previous section described the manner in which any form of leverage based on changing the proportion of fixed costs affects risk. Since debt carries the fixed expenses of interest and the repayment of a principal (unlike equity), more financial gearing means more risk. We have, in order to be confident about using our competitor's β, to un-gear it – to remove the influence of the competitor's chosen level of financial gearing. This is done by dividing by the quantity (1 + *debt-equity ratio*), although this is something of an approximation to basic principles.[9] Assuming that the competitor's debt-equity ratio is 20% then, conveniently, the company β that we use is 1.2/120%, or 1.0.

We now need to adjust this β for the particular characteristics of the proposed project. Suppose that the benefits of the project are based on improving revenues. Perhaps they are to do with reducing the size of product delivery periods and with increasing their dependability. This means that the first risk factor, the revenue sensitivity, is not zero (as it might be in a fixed cost-saving project) because the benefits will disappear if nobody buys the firm's goods. Suppose however that the feeling is that the effects of general economic conditions on the revenue benefits will be less than on the company's turnover as a whole. This might be because the improvements in lead-times will make the demand for the associated products more constant, as well as larger. In particular, customers are happier about ordering goods with short lifetimes when economic conditions are rather changeable. It means *they* carry less risk that they will be stuck with components which they are unable to convert into finished goods or final services.

We shall assume that the sensitivity of the project's turnover is a half of that of the company as a whole, and that the first risk factor (f_1) is therefore 0.5. In other words, we are predicting that if the company's turnover drops by 20% in aggregate, our benefits (which are synonymous with revenues here) will only drop by 10%, to £540000.

The second factor (f_2) is concerned with the sensitivity of the project's returns (that is, its NPV) to changes in these revenues. A crude way of finding this (in keeping with a desire to have things as simple and transparent as possible) is to calculate the sensitivity of returns to revenues at a fixed point. We will use the point at which revenues change by −10%. To do this, we shall have to *assume* a value for the discount rate, work out the sensitivity, and then apply the sensitivity to find the real discount rate. We have, in other words, to work iteratively. We shall in fact base the assumed rate on the β of the whole company which, as we have already seen, is exactly one. Taking the risk-free rate as 10%, and the market premium as 6% (the historical average also mentioned in the last section), the temporary discount rate is

$$\begin{aligned} k &= R_f + \beta(R_m - R_f) \\ &= 10\% + 1 \times (6\%) \end{aligned}$$

or 16%. With the predicted cash flows, this gives an NPV (in £000s) of

$$\begin{aligned} V &= \sum_i C_i/(1+k)^i \\ &= -1000 + 500/(1+0.16) + 500/(1+0.16)^2 + 500/(1+0.16)^3 \\ &\quad + 500/(1+0.16)^4 \\ &= 399 \end{aligned}$$

If we vary the revenues downward by 10%, the cash flow schedule changes by reducing the benefits to £540000, and therefore the net flows in Years 1 to 4 become £440000. So the NPV is now

$$\begin{aligned} &-1000 + 440/(1+0.16) + 440/(1+0.16)^2 + 440/(1+0.16)^3 \\ &\quad + 440/(1+0.16)^4 \\ &\quad = 231 \end{aligned}$$

This is a proportional change of (399 − 231)/(399), or 42%. It is much higher than the proportional change that induced it, the 10% fall in revenues, because there are no variable costs: it was assumed that the project's costs were unchanged by the fact that the firm sold fewer products. The high gearing that follows is probably quite representative of real projects of this type. The sensitivity of returns to revenues is therefore 42%/10%, or 4.2. We need, however, to re-express it as a quantity relative to that of the company as a whole. It will be assumed, once again very conveniently, that the sensitivity of the company's returns to its turnover is 2.1; (in other words, that if turnover drops by 10% then profits will drop by 21%). The second risk factor is therefore 4.2/2.1, or just 2. Finally, then, the project's β is the company β multiplied by the two risk factors: that is

$$\beta_p = f_1 f_2 \beta_c$$
$$= 0.5 \times 2 \times 1 = 1$$

This makes the project discount rate $k = R_f + \beta(R_m - R_f)$ exactly 16%. At this rate, we have already seen that the NPV is £399000, and this, therefore, is the economic worth of the project.

It is of course conventional to lay such calculations out in a neat, tabular form, and to relegate the narrative to an appendix, since it otherwise detracts from the sparse precision of the figures. The words are, all the same, an integral part of the calculation, and it wouldn't be sensible to separate one from the other. It is in fact important to say much more than has been said here. In particular, one has to recognize that net present values tend towards a value in equilibrium of zero under the action of competition. There has to be a good reason why present values are anything other than zero – a competitive advantage of some kind, or a short-run deviation from competitive equilibrium due to some unforeseen event.[10] It is worth spelling out exactly what the source of present value is supposed to be, because all else springs from this. The understanding comes first, the numbers later.

Changes to the discount rate

Much of the manipulation with the discount rate disappears in projects that derive their benefits from saving fixed costs. For the reasons already discussed, the first project risk factor is then zero, and so, therefore, is the project β. This means that the discount rate is simply the risk-free rate of interest, or 10% in our example. In many developments concerned with engineering and manufacturing this is substantially the case, and calculations are therefore much simpler. But one has to remember that the newer technologies have had a disappointing effect on the cost base of companies that have adopted them, and, as explained in Chapter 3, we should also be looking for gains on the revenue side. This being so, it will not always be acceptable to take β to be zero.

The next simplest case is one in which the risk of a project is close to that of the company's operations as a whole, and the adjustment to β can then be taken to be unity. As will be seen in Section 5.3, the two risk factors f_1 and f_2 tend to offset one another for computer systems, and it often seems to be quite sensible to use a β adjustment of one. (In other words, to avoid adjusting the company β at all.) But even if this were not the case, it is not really very taxing to find a specific discount rate provided that a certain sense of perspective is maintained. It is clearly difficult to be precise about quantities such as the sensitivity of revenues to fluctuations in general economic conditions, but it is not so difficult to express general feelings to within an appropriate ten or twenty per cent.

As a matter of interest, we can see how changes in the discount rate affect the final NPV. Suppose that, in our example, we took the revenue sensitivity to be the *same* as the value for the company as a whole, instead of half of it. The project β would then be 2 and the discount rate 10% + (2 × 6%), or 22% instead of 16%. The NPV is then

$$-1000 + 500/(1 + 0.22) + 500/(1 + 0.22)^2 + 500/(1 + 0.22)^3 + 500/(1 + 0.22)^4$$
$$= 247$$

(that is £247000). This is less than two-thirds of the earlier NPV of £399000, and the difference is evidently substantial. On the other hand, if the project had been directed towards saving fixed costs, β would be zero, the discount rate just 10%, and the NPV £585000. I think this indicates the value of spending a little effort considering the effects of non-diversifiable risk, the two adjustment factors, and the appropriate company β.

Some omissions

There are two elements that have been ignored here. The first is price inflation. It is generally held that interest rates have a component equal to the expected rate of inflation, and that the remainder, once this has been deducted, represents a so-called real rate. This is not an entirely accepted principle, and if we were to explore the issue in detail we would need to start worrying about how interest rates varied for capital instruments maturing at different times. But the discount rate that we have used *is* typically formed in inflationary conditions. It would therefore be wrong to attempt to express cash flows in so-called real terms. They must be stated in nominal terms. Suppose, for instance, that we expected to gain revenues in Year 4 of some £700000 rather than £600000, as a result of a general price inflation. Then we should use the £700000 in our calculations. It should not be adjusted downward to reflect a notional, underlying value.

Of course inflation is one of the many influences on cash flows, and the manner in which it takes effect may be relevant to the success of a project. There may, for instance, be a difference between the rates at which prices of raw materials, and those of finished goods, rise. This means that the attractiveness of a proposal could depend on the assumptions made about the pattern of future inflation. It is not, however, especially relevant to the appraisal of new technology.

The second missing element, taxation, has been ignored because its treatment hinges so much on local circumstances. It depends not only on the particular regime in force in a specific country at a specific time for a specific type of corporation, but also on such issues as whether accelerated depreciation allowances are available and whether the company has tax losses.

It should be included in the calculation as an additional set of cash flows representing exactly that amount of additional tax that the company will pay on its increased earnings – the earnings that flow from the new development. One has to be careful here because taxes are based on accounting earnings rather than the cash flows used in the NPV calculation, and, to make matters messier still, these accounting earnings differ in certain respects (such as the treatment of depreciation) from a company's financial accounts. Taxation is plainly a significant

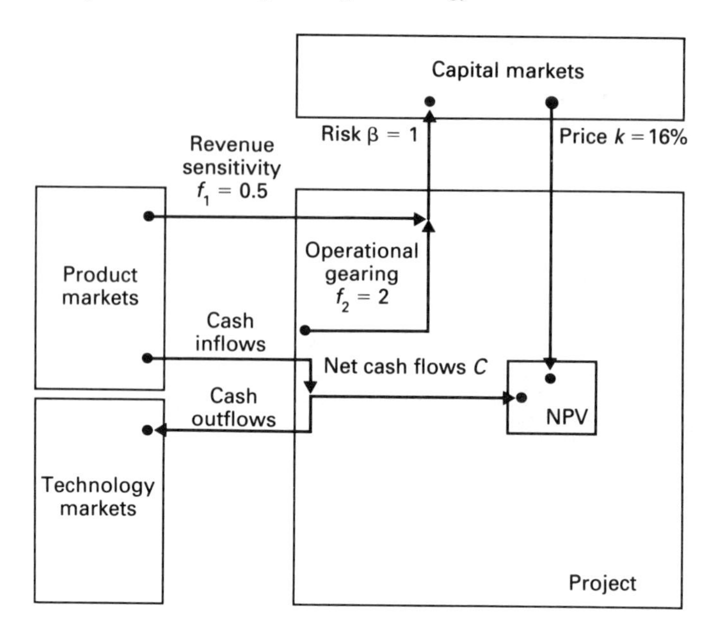

Figure 5.4 A project's present value

influence on the way a company conducts its financing activities, but it would be wrong to become pre-occupied with it when managing operating activities. To change the nature of a firm's manufacturing systems (as opposed to its financial structure) for tax reasons is to risk diverting attention from more central issues.

Figure 5.4 provides a summary of the example described in this section. From what little we know about it, the development in this example appears to be a highly worthwhile one. Of course in reality a company cannot make decisions about individual opportunities without regard to the systems it already operates and any other opportunities that have presented themselves at the same time. It makes sense to pursue a coherent strategy – a pattern of developments which will build one on another over the subsequent years. It seems unacceptable to evaluate individual opportunities as though they were entirely unconnected.

Fortunately, this interconnectedness neither invalidates nor contradicts the application of the present value yardstick to specific proposals. It is perfectly consistent to have a plan within which proposals for new systems are generated, and a way of then assessing them individually. We might expect, in fact, that proposals that stem from the strategy will have a greater present value than those that don't. The fact that the strategic fit is good suggests that their connections with other developments will be sound and straightforward, and that those who work on them will do so with a sense of direction. It is plainly important that a firm can take advantage of existing expertise and that it can integrate the operation of the different systems it adopts. This is really not feasible without some kind of strategy. A development adopted by a directionless organization operating uncoordinated systems will probably have relatively little present value.

However, if we wanted to observe the value maximization rule, our new opportunities for investment should *not* go through a filtering stage in which they could be dismissed on grounds of poor strategic fit. In such an instance there is a danger that what might prove to be very successful is rejected on specious grounds. So we ought to be wary of making a strategy so rigid either that unexpected opportunities will be dismissed before they are evaluated, or that the individual elements of the strategy will not be tested on the basics of their own distinct costs and benefits. This suggests that strategies are consistent with maximizing value only if they are applied in a relatively weak form – if they are used to guide the search for new opportunities, but not their evaluation. There will be more about forming and following strategies for new technology in Chapter 7.

5.3 Computers and cash flows

There are two components of the net present value yardstick that are, at least partly, within the control of people working on a particular development: the pattern of cash flows that it is expected to generate, and the relative sensitivity of its returns to changing economic conditions (its non-diversifiable risk). In the latter two sections of this chapter, our concern is with issues that affect the practical application of the yardstick to computer-based manufacturing systems, and these two components will be discussed in turn.

Patterns of cause and effect

It has already been mentioned in Chapter 3 that the operating effects of advanced technologies are often several stages removed from the financial consequences. One has to pursue an extended line of argument over a chain of several causes and effects to determine the measureable impact of implementing a new system. This is particularly so when a firm is attempting to boost revenues, for any improvement in process technology is bound to have only an indirect effect on the commercial conditions in the firm's product markets. The problem is to connect technologial properties (such as flexibility) with the impact these have on the conduct of a firm's operations: and then to connect the operational effects (such as faster product introduction) with the prices and volumes of output the firm can sell to its customers.

This suggests that there is a set of intermediate variables whose influence (and the extent to which they are influenced) determines the financial impact of a new technology. Suppose, for instance, that a firm is considering how it might improve the information flows that occur just in advance of its production activities – the passage of product drawings, process plans, production schedules and so forth to shop offices and cell controllers. It might in fact want to automate entirely the transcription of process plans and schedules from, say, the MRP database to an FMS controller.

It will doubtless expect to observe many fewer errors in the plans and schedules at the FMS once manual transcription has been eliminated. It might also expect improvements in timeliness, since there should be little delay once the passage of information is automated. These effects will eventually influence measurable, financial quantities. But the scale of these effects is only understandable in terms

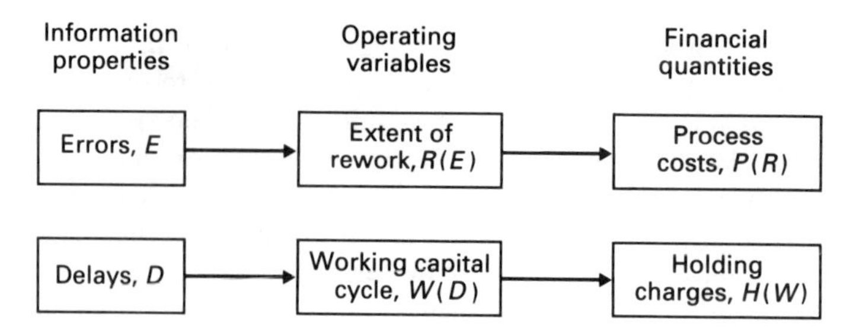

Figure 5.5 Intermediate variables

of intermediate variables such as the scale of rework that is saved by a reduction in errors, and the fall in work-in-process that follows a reduction in the delays associated with information delivery. Figure 5.5 indicates that rework is a function of errors, and that costs are in turn a function of rework.

For example, we might expect that errors will occur during manual transcription at the rate of one for every distinct object transcribed (every process plan or schedule, that is). We might then estimate that one in ten of these will be severe enough to cause a product to be eventually reworked. If the cost of rework is roughly £1 000 per product, and the rate of production is 1000 products each year, the annual cost incurred as a result of manual transcription will be roughly £100000. This represents the limit of the savings we could make with automation (ignoring other kinds of effect). Without explicitly thinking about an intermediate quantity like rework in this manner, it would be very hard to work out how costs depended on errors.

Unfortunately, an influence that on first appearances might act through a linear chain of causes and effects often becomes a lot less tidy on closer examination. An FMS on its own, for instance, is not enough to bring down inventories. A firm may have to re-organize its production systems to take advantage of its slick changeover, handling and transport mechanisms. It may have to re-design its product to make sure it can be properly manipulated. And it may end up simply shifting a bottle-neck somewhere else – with nothing but a marginal improvement in lead-times.

The cause-effect pattern is also complicated by the fact that elements like reduced lead-times can be traced to more than one type of commercial benefit, each providing a positive cash flow. In addition to helping reduce inventories, smaller lead-times reduce the likelihood, and therefore the expected cost, of delivery penalties. They probably increase revenues by making the firm's promised delivery performance more attractive to potential customers.

In other words, the chains of cause and effect are in fact linked to form networks. Figure 5.6 is meant to suggest this by showing how the earlier example of automated information transfers might be analysed. A given cause can have more than one effect, and an effect more than one cause. If an effect has more than one cause, it will sometimes be necessary that all causes occur in combination if the effect is to be experienced: sometimes any one of the causes on its own will be enough. Because new technologies exhibit such an extended pattern of causes and effects, we might

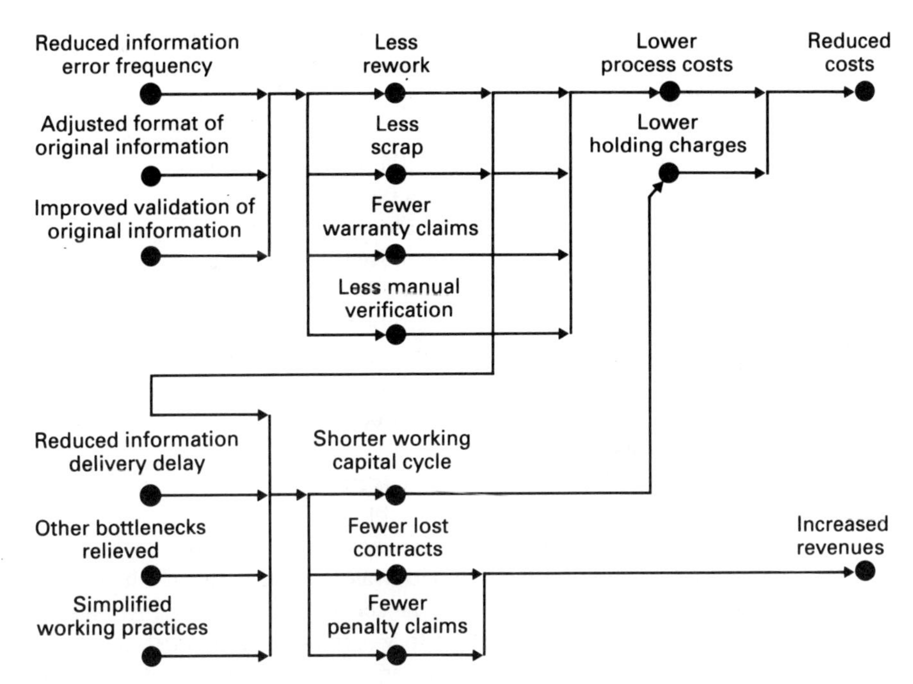

Figure 5.6 Cause and effect networks

say that their appraisal comes up against the problem of *diffuseness*.

One way of tackling this complexity would be to hold everything constant other than a single technological variable and to assess the financial results. This would allow us to analyse an extended pattern of causes and effects, essentially limiting its scope to just one causal factor. But this is a very restrictive manipulation, for it ignores the need to change many elements together. In particular, we know it is not true that with computers the best an organization can do is tinker with the old order – that it can only do what it has done before with different mechanisms. The implication is that we cannot turn our cause-effect network into a simple chain by keeping almost everything fixed while we consider the influence of technological issues alone.

There are, however, a number of difficulties involved in manipulating a cause-effect network in its entirety. The first is simply that it is hard to gather the information needed to build it – to make sure that every connection that could possibly be relevant is included, and that the strengths and directions of causation from one element to another are known. If the cause-effect network is to be used as a way of *calculating* rather than judging the scale of a technology's cash flows then we would have to quantify the influence of each element on every other element. We would have to be able to gauge, for instance, how many extra degrees of flexibility we gained from introducing a new manufacturing cell, and then how sensitive product introduction times were to each additional degree of flexibility. This would mean having to find a way of quantifying flexibility, as well as a way of quantifying the connections of other elements with it. In fact it was said at the outset (in Chapter 1) that it is usually unrealistic to assume that the relation between the inputs and outputs of an organizational process is deterministic. In other words, no matter how much information we gathered, we would never be able to say for sure what the consequence of taking a given action would be.

A second problem is that the accumulation of random errors you would observe as a network of this kind is traversed makes the output very fuzzy, even if it is possible to be precise about the inputs. Estimates of the impact of an FMS on lead-times might be uncertain by 30%; estimates of the impact of lead-times on inventory levels uncertain by 20%, and so forth. By the time it gets to establishing the impact of the decision (to proceed with an FMS) on the yardstick (the scale of positive cash flows) the analysis may be getting hopelessly vague.

A related difficulty stems from the accumulation of consistent error. This has the effect of making the final outputs highly biased, rather than uncertain in a random sense. This need not of course be deliberate: a sustained optimism is enough to yield wholly misleading results. The particular difficulty here is that, by breaking down the analysis into small elements, it is easy to accumulate distortions without them ever becoming noticeable.

The third problem is that the effects do not act instantaneously – there is a delay and a period of initial, unsettled behaviour associated with each link in the network. A given set of causes will yield a given set of effects that change in scale with the passage of time. We have therefore to worry at least as much about the dynamics of technical changes as about any equilibrium that might finally be achieved. This means understanding the delays that will be experienced before cash flows are observed, and even perhaps whether the process of introducing a new system is a stable one. It is conceivable that, if a development is managed in a reactive, implusive fashion, its operation will only ever swing from one unhappy extreme to another.

A final difficulty with analytical solutions to this network is the complexity of the manipulation involved. Once we start to introduce uncertainty and delay, even quite simple patterns of cause and effect yield some forbidding expressions for the relationship between the final financial effects and the initial technological causes. To get a flavour of the mathematics you might want to look at some of the literature that has emerged from the field of information economics.[11]

Drawing networks of causes and effects nevertheless remains a good discipline. It is a way of sharpening up one's thinking about the consequences of initiating new developments, and of the associated work on which their success depends. It brings certain assumptions out into the open where they can be verified or corrected. It helps reinforce the analyst's understanding of symptoms and diseases: in what respect, for instance, piles of work-in-process are a signal that there are more basic problems (such as machine breakdowns), and in what respect they are themselves a source of problems (such as holding charges). The current emphasis on practices such as activity-based accounting, in which one is looking for the most influential determinants of cost, illustrates that when conditions and systems change, old rules of thumb for working out financial consequences become obsolete. The correlation that was apparent in the past breaks down and one has to go back to examine causation more thoroughly.

When firms apply a technology that is new to them, it seems to be common that they forget about antecedents (all the causes that precede an effect), and they ignore consequents (all the effects that follow a given cause). They fail to consider such essential antecedents as technological education and training, with the result that the intended benefits of a development fail to materialize. And they don't bother to think about consequents (particularly beneficial ones) which are unfamiliar or ill-defined, with the result that valuable projects sometimes never get off the ground.

So a reasonable conclusion to draw is that an understanding of the pattern of causes and effects is a worthwhile component in the process of finding cash flows. But in most instances it will not be possible to quantify the internal elements of this pattern and perform calculations on it. The best that can probably be done is to use the pattern's qualitative form as a basis for subjective judgement. In other words, given the outlines of this pattern, the analyst must simply judge the scale of positive and negative cash flows that arise from that initial causes over which the firm has influence.

This is unsatisfactory in some ways, because it means that we have an analytical model for the connection between cash flows and present value, but not for the connection between technological actions and cash flows. This is, however, what we might expect, because once we have a picture of the cash flows we have a highly abstract picture that is not, in fact, very difficult to manipulate. The hard part is cutting away the complexity of the real world in order to get to such an abstraction in the first place.

Prediction is in other words more a process of judgement than calculation, and this being the case it is important that the judgement is an informed one. The role of an effects network is to make sure that this is so.

Incorporating non-financial transactions

Some activities that people and firms take part in do not, in any obvious way, involve finance. Some of the exchanges that are made, even though they enhance the well-being of those entering into them, do not go through a monetary stage. Not only do the participants find it unnecessary to use money as a means of exchange, but they frequently feel no great need to account for that exchange in monetary units. When, for example, a manager invests some time in training a subordinate, neither one nor the other pays any money, nor is the training ordinarily recorded as a monetary transaction in a ledger. This lack of financial visibility does not, of course, imply that there is no economic substance to the training: it makes the trainee better off by increasing his stock of knowledge, and it probably makes the company better off by increasing the quality and quantity of the trainee's work.

So while we can be reasonably certain that economic arguments encompass all the issues that need airing during appraisal, we can't be sure that a financial view is wide enough. Whereas we want to capture a notion of worth and sacrifice in the broadest sense, the finance discipline is focused quite closely on resources that exist in monetary form, or that have at some point been exchanged for money. We would obviously like to be able to translate opportunity costs and benefits of the sort acquired in non-monetary exchanges into financial quantities. Present value calculations can then be performed. But we cannot be sure that this translation will be an entirely accurate one. Our main worry is that the financial view of a new technology's unusual and diffuse benefits will be understated. This is especially likely when people carry over the accounting frame of mind to appraisal because prudence and objectivity are among the criteria by which the goodness of accounting information is judged. Prudence is a euphemism for a consistent pessimistic bias, and objectivity one for ignoring everything that has no documented history.

It is normal to refer to the problem of assessing influences that cannot, in practice, be quantified in monetary units as being one of *intangibility*. This can be misleading if it is taken out of context: one can also use the term to describe imaginary rather than real phenomena. The danger is plainly that a decision maker might automatically assume that a factor characterized in this way will never give rise to observable changes in a firm's performance. Here we are specifically concerned with factors that do have substance but which, for one of the reasons just mentioned, are not readily connected with immediate cash flows.

This suggests that a firm should take the process of examining financial effects as far as possible – to a point, perhaps, at which its analysts begin to lose all feeling for the uncertainty associated with the values they are predicting. But beyond this, any residual sources of cost and benefit ought *not* to be ignored, since to do so would be to misrepresent the broader balance between what must be sacrificed to introduce a development and what will be gained from it. It may be that any net, economic worth will finally be manifest in financial flows – bigger dividends for shareholders perhaps – but there are additional mechanisms needed to turn such worth into financial flows. And it is a lack of knowledge about these mechanisms that causes us to leave the worth unquantified in the first place.

The most obvious way of coping with issues that firmly resist expression in financial terms is to postpone their consideration until the elements that *can* be expressed financially have been taken care of. If a financial analysis indicates, on its own, that a development is worth pursuing, then it may be felt that it should be adopted without worrying too much about benefits that have not been quantified. If, on the other hand, the financial analysis yields a negative net present value, those making the decision have a single present value to weigh against the unquantified benefits. Should the decision makers' judgement suggest that the unquantified benefits count for more than the negative present value then they would presumably proceed: otherwise they would take the proposal no further.[12] We might, for example, find an NPV for a proposed desktop publishing system of *minus* £50000. If this didn't take into consideration our expectations of enhancing the firm's sales by providing more appealing pre-sales specifications (an effect we are unable to quantify), we would simply need to ask whether it is worth paying out a present value of £50000 in order to secure that benefit.

This is a case, in other words, of reconciling costs and benefits, or present value and intangible effects, by the application of ordinal rather than cardinal scales. We may not be able to speak about the magnitude of the difference between one effect and another, but we ought to be able to estimate its sign – to say whether it is positive or negative. As with the diffuseness problem, the final decision relies on a subjective choice rather than the outcome of a strictly logical calculation. Once again, our objective is to make sure that this judgement is a properly informed one.

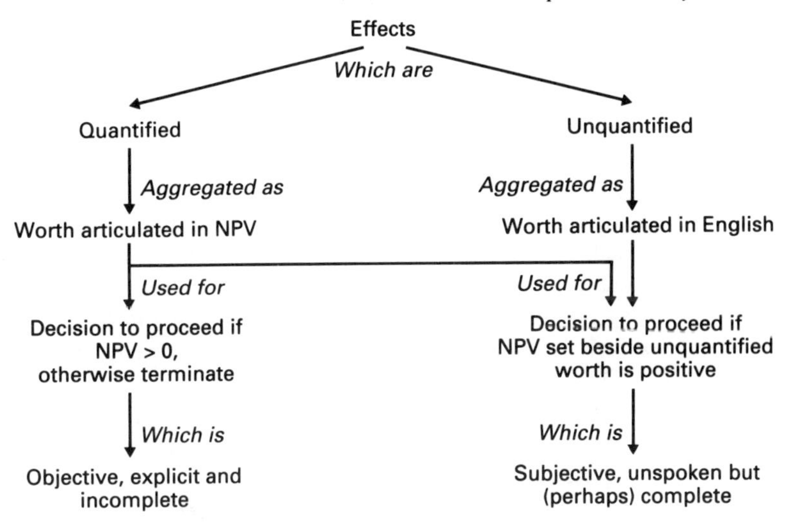

Figure 5.7 Coping with unquantified information

This process has its disadvantages, for without a quantified figure of worth to encapsulate *all* the effects of a development it will be harder to assess whether, during its introduction, it is meeting expectations. But at least it narrows down the information that a decision maker has somehow to keep in his mind's eye: a financial value on the one hand, and a qualitative statement of an additional benefit on the other. Instead of reconciling different effects in an explicit manner by reducing them to degrees on a common scale (a financial present value), the final balance is assessed in someone's head – or in a committee's processes of reconciliation and consensus. That suggests a loss of objectivity, and hence the need to narrow down the scope of this process as far as possible by making it a comparison of just two elements. But it is done in the recognition that the next best solution – a purely quantitative process – omits significant elements. Figure 5.7 summarizes these points.

There is a way in which we can, apparently, avoid the kind of direct analysis which has just been discussed. This is to say that the scale of a new system's benefit is equal to the saving made by applying an advanced technology in place of a conventional technology which would achieve precisely the same ends. Typically, the advanced technology is flexible and automated, while the conventional type is inflexible and mechanized. If we were assessing the value of an FMS, we would attempt to find the cost of introducing the stand-alone machine tools needed to produce the variety and throughput of which the FMS was capable. We would then subtract the cost of the FMS from this and be left with the saving due to the FMS.

Of course this approach assumes that the only alternative to an advanced system is a new conventional system with precisely the same capabilities, and this is *never* true. The risk one takes in applying such a line of reasoning is that it only compares technical solutions to given problems – it plainly doesn't evaluate the problems themselves. It would not, for instance, reveal whether there is any benefit in producing the variety of products that an FMS can in principle turn out.

Finding opportunity costs

It was said earlier that the costs included in an appraisal are relevant opportunity costs expressed as cash flows. To be relevant they must be attributable to the particular decision being considered: to be opportunity costs they must express the benefit forgone by taking one course of action in place of another: to be cash costs they must not have undergone any treatment such as accrual or apportionment.

An interesting case, also mentioned earlier, is that of work that uses the services of internal departments – of people employed by the company on unspecific contracts. (In other words, of people who are paid a salary in exchange for an undertaking to carry out unspecified duties.) Such a service might be the writing of software which, because of its specialized nature, cannot be bought as a proprietary system. The approach sometimes taken is to evaluate the feasibility of projects using a standard recovery rate for software development. This recovery rate is a way of dividing up the software department's expenses and attaching them to the projects it undertakes according to the number of man-hours devoted to each project.

Although this approach was taken in the example of the previous section, it is not a completely satisfactory one. Recovery rates provide a way of apportioning costs that would be incurred whether a project went ahead or not. The same salaries are paid, the same heating bills are paid, and many of the same maintenance expenses are incurred irrespective of a single project's fate. In other words, salaries, heating bills and maintenance costs are irrelevant to the viability of the project: they are sunk costs – they cannot be undone. Recoveries do not satisfy the tests we want to apply to the costs we incorporate in the NPV calculation. But there *are* opportunity costs: if by working on project *X* the development team can no longer implement system *Y*, then the cost is that of losing the benefit of system *Y*.

The fact that sunk costs are out, and opportunity costs in, presents a difficulty. Sunk costs are relatively easy to measure because they have already been incurred or committed and, presumably, they have been documented. Opportunity costs express the benefit that is forgone by not spending resources on doing something else, and they are therefore much harder to establish. In estimating these theoretical benefits, all the difficulties of diffuseness and intangibility discussed earlier have to be faced again. It is fortunate that this is only a problem with resources obtained from within the company: something bought from outside has an opportunity cost equal to its price on an open market.

At the end of the day, of course, a business has to be pragmatic about these matters and it isn't obvious that there is a better way in which the costs of software development can be expressed. It would be impractical, each time a company began to evaluate a new project, to look at every alternative opportunity and establish its value – simply to determine the benefit that would have to be sacrificed by ignoring it. There is no guarantee that all the other opportunities (some of which might not be evident until slightly later) *could* be identified, let alone identified at a reasonable cost. It would be a lot simpler just to work out an hourly rate by which development hours are multiplied to find the total cost. This rate might be changed from time to time, but not every time a new opportunity arose.

But how such a rate should be determined is hard to say. In fact, to come full circle, you might argue that this opportunity cost rate is probably close to the recovery rate. You might say that, in an organization of moderate efficiency, the benefits of the second-best projects are probably going to be somewhere close to the expenses of running the development department. This is not *necessarily* the case, but it may be a reasonable assumption. It may be more reasonable than spending money to employ more investment analysts.

The argument is really that making rational decisions involves making a judgement about costs. It isn't defensible to follow blindly a rule about recovery rates, although in specific circumstances these rates may be applicable. For there may well be occasions when forgone benefits are substantially greater than simple recovery rates. In fact, one of the advantages of consciously considering opportunity costs is that they force an analyst to consider alternatives. He cannot speak about the sacrifice of forgoing the next best option without exploring the options that are available to his firm.

5.4 Computers and risk

Risk gains

Non-diversifiable risk is important because it determines the discount rate (the rate at which future cash flows must be reduced to obtain their present values). In the previous section it was seen that there are two main factors that influence this risk: the way in which a particular investment affects the volatility of a firm's revenues, and the way in which it affects the relative size of fixed expenses. The greater the sensitivity of revenues to fluctuating economic conditions, and the greater the proportion of fixed costs, the greater the risk borne by the firm.

Considering risk is also important for more general reasons. Even to managers who wish to know nothing about net present value. Most people, administrators especially, would like to reduce the amount of uncertainty and variability in their working lives: they would like to be able to predict future conditions perfectly, and to understand the degree of uncertainty they face when it cannot be wholly mitigated. This aversion to risk is implicit in the net present value calculation, although it is the risk aversion of shareholders that underlies the theory. Risk seems to be as big an issue in the lives of industrialists as it is in those of financiers, even if it is characteristic of industrial risks that they are less easily measurable and less explicit than the financial kind. It is, all the same, worth bearing in mind that the concern here is not with all the consequences of uncertainty: only with non-diversifiable risk. This means that measuring non-diversifiable risk does not, necessarily, reveal enough about uncertainty for good project management. The practical analysis of risk is described in Chapter 8.

To consider the first factor then, how might computer-based technologies affect the sensitivity of revenues to general economic conditions? Could these technologies make a manufacturer's turnover more or less sensitive to what takes place beyond the factory gates?

For the most part, the sensitivity of a market's size to changes in national incomes, unemployment, price inflation and so forth is beyond the influence of the firm. If customers in that market postpone or cancel their spending when confidence is low there is little that a supplier can do. There might, in fact, be some marginal effects: better quality products or faster deliveries might, in reducing the uncertainty faced by a customer when he makes a purchase, make him less pessimistic when times are hard. But other than this, a firm's only way of compensating for a shrinking market is to increase its market share – to exploit its competitive advantages more intensively.

This means that any of the revenue benefits discussed in Chapter 3 can reduce revenue sensitivity on a transient basis. But the sustained ability to cope with economic fluctuations is somewhat different. The key is to recognize that the nature of demand in most markets is always changing, and that to reduce revenue sensitivity a firm needs to display a measure of adaptability – a capacity to track changes in the product attributes that customers prefer. When low prices assume a greater importance than a product's form or functional elements a supplier needs to be able to reduce his cost base. When customers want to run down their stocks of incoming commodities the supplier must think of reducing the lead-times and batch sizes of the outputs he provides. A responsiveness to customers' desires will reduce a supplier's vulnerability to conditions at large if it offsets his customers' sensitivity to those conditions at large.

Computer-based technologies can help in various ways. First, computers are good at collecting information and identifying patterns in it. They can help manufacturers understand what is happening to the demands for their products, and they can show how current conditions compare with those of various periods in the past. They can simulate the ability of a production operation to meet changes in demand. That is, they can help predict the effects of impending changes in the first place. The extent to which this is feasible depends on the industry, but the idea is that this sort of technology can help produce a good deal of the information that businesses find useful in planning how to cope with varying economic conditions.

Second, automation can help firms deal with the inertia of mechanization. Mechanized plants tend to be expensive and inflexible: product introduction and product changeover is costly and time-consuming. This makes it hard for companies to adapt to changing patterns of demand – whether for different types of products, shorter lead-times or smaller batch sizes – and it makes their turnover more sensitive to these changing demands. Many of the newer types of factory automation are directed less at reduced material and energy consumption (and people usage) than at flexibility. They are meant to be capable of producing products of changing form and function. They are meant to minimize the cost penalties associated with making small batches to short lead-times.

Of course one can overstate the contribution of computers to this type of effect. In most instances, computers are a necessary but far from sufficient element. They have made programmable plant such as robots and CNC machines possible and they have made quick, adaptable scheduling possible. They have raised the prospect of distributing product and process information throughout the factory quickly and accurately. But if at the end of the day people do not exercise sufficient care and competence to exploit flexibility, automation will have little material impact. Flexible systems can be used in remarkably rigid ways.

So, third, we come to areas where new technologies are beginning to help firms cope with the inertia of organization. Here, the purpose is to improve the effectiveness of processes in which people participate: managing contracts, designing products, designing production tools, planning production schedules and so forth. Any way in which a system makes adaptation less costly or less time-consuming will help reduce risk.

A number of office-based systems, for example, are directed towards improving the ability of their users to get products to market. These include project support environments and engineering data management applications. The idea is to help manage the *process* of engineering, and an important element is the ability to take charge of the vast amounts of information generated and consumed in and around a firm's engineering departments. These technologies make information (such as part drawings) visible and accessible, while ensuring that its availability is restricted to the right people at the right times. They formalize communications by polling and notifying people when a new component has been designed or an old component changed. And they promote re-use by keeping track of what's been done in the past.

Without descending to a further level of detail, it ought be apparent that this sort of technology helps firms adapt quickly to changing external conditions, particularly demands for new or modified products. Systems of this sort provide an infrastructure for slick communications and information management during the stage of the product cycle where substantial lead-times are consumed, and where the vast majority of the product's costs are determined (if not incurred). More general schemes of this sort (supporting office-based activities other than engineering) are emerging from developments in the area of computer-supported co-operative work.

An argument sometimes raised against systems of this kind is that they operate in a much more rigid manner than people, and that they make little sense when most firms are looking for less rigid ways of conducting their operations. The point of using such systems, however, is only to automate the routine elements of procedure – the elements that are repetitive, simple and often very irksome. They relieve people of the need to talk to one another about the mechanisms of bureaucracy, and give them the chance to talk about more challenging issues. It is also as well to remember that one can re-program a computer in a way that one cannot re-program an organization. If it is the computer alone that has to remember and act upon mailing lists and vetting lists and approval boards, then it is probably much easier to change the procedures that affect them than it would be if a hundred people had to remember and act upon the same things.

As ever, the scale of the influence on the discount rate is hard to quantify. The question to be asked is how much more or less sensitive to aggregate economic changes are the revenues associated with a particular project than those of the firm. We know that at one extreme, when the project is intended only to save fixed costs, the sensitivity is zero (because, by definition, fixed costs are independent of revenues). We also know that a project whose revenues are no more or less vulnerable than the firm's revenues in total has a sensitivity factor of unity. This suggests that if we have no information whatever it would be sensible to put f_1 equal to one.

Placing a specific development at some point in the gaps is then a matter of judgement – a judgement that is perhaps more a commercial one than a technological one. It is a case of identifying the commercial effects of a new development (that is, anything that affects the level of demand in the firm's product markets), and asking people with a reasonable understanding of the product markets whether they think these effects will make sales more or less robust than formerly. For example, if we are pursuing a development that will reduce lead-times by a third, we would doubtless ask the firm's salesmen how this would affect revenues. This would reveal the positive cash flows of the project. We would then need to ask them whether they thought that improved lead-times would make customers less willing to stop buying our products when conditions in the world at large changed for the worse. If they thought that this were true, we would at least know that the revenue sensitivity lies between zero and one.

Risk losses

Although new technologies can help reduce non-diversifiable risk in some ways, they almost certainly increase it in others. This is a result of the way they influence the second factor that was mentioned when discussing risk – operational gearing. The argument was that the higher the proportion of a firm's operating costs that were truly fixed then the more sensitive were its earnings to ups and downs in turnover. The reason was the scissors-effect between costs and revenues sketched out in Section 5.1.

The plain fact is that new systems incur few costs that vary with the volume of business that the company conducts. Production systems, like flexible manufacturing cells, may have such varying costs in the raw materials that are formed into its products; and they may have appreciable variable expenses associated with the energy they consume. But they will generally have a greater fixed cost base than the systems they replace. Technologies whose main, or only component, is a computer will have very few variable costs indeed: it is only some of the consumables (like certain storage media, perhaps) that vary with the extent to which the system is used. Even this is a stage removed from variability with turnover, because the extent to which a computer system is used – in terms of storage and processing time – isn't necessarily associated with the strength of the company's order book.

Instead, almost all costs are as good as fixed. Buying a system, developing application programs, providing an infrastructure, operating and maintaining it are largely determined at the beginning of the system's life. Decisions that may be made or reversed at an intermediate point in a system's life, like developing a new application, make no difference: the costs of that additional application are still fixed for *its* life. They almost certainly cannot be continuously varied with the company's turnover. So the company is stuck with these fixed costs, in the short-ish term, whether business is good or business is bad.

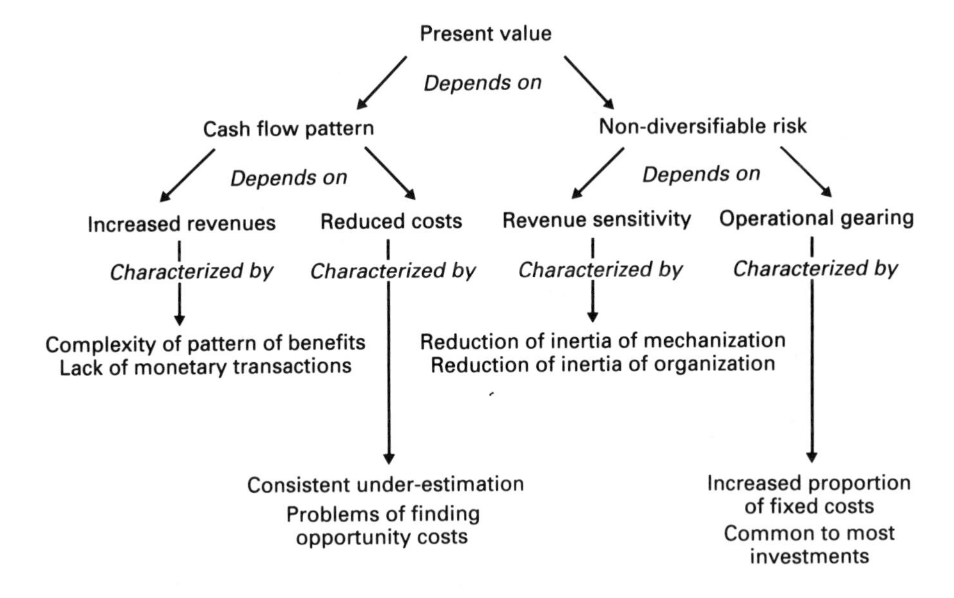

Figure 5.8 The effects of advanced systems on project-dependent factors

The result of fixed costs being a high proportion of total costs is, by definition, a high level of operational gearing, and a relatively high level of risk. This is not to say that the level of risk becomes unacceptable, just that it influences the net worth of a development adversely. In any case, high fixed costs are the case with just about all new investments, including those in organizational changes.

Figure 5.8 summarizes this, and the other general properties of new technology in terms of the components of present value that vary between one development and another.

5.5 Summary

A financial appraisal is used to make a prediction of how big a surplus a firm can earn by applying capital in a particular way. We can find the opportunity cost of this capital by looking at highly developed financial markets, and by observing the price they set for capital applied to investments with a specific set of characteristics. It turns out that this price depends on a pure time value and on the level of an investment's non-diversifiable risk – the element of risk (due to conditions in the economy at large) that cannot be reduced by holding portfolios of assets whose returns tend to move in opposite directions.

Time value is easy to observe: it is the interest rate on capital that is effectively risk-free (and government securities the obvious examples). The risk premium is a straightforward, linear function of non-diversifiable risk. But non-diversifiable risk itself is hard to reason about: for complete firms we can simply look up values (on a special scale) in books, but this must be adjusted to reflect the characteristics of specific projects. We do this by finding the relative sensitivity of the project's returns to general economic fluctuations. This is split into a revenue sensitivity (the rate at which revenues change with general fluctuations) and an operational gearing (the rate at which earnings change with revenues due to the scale of fixed costs).

Once we have a price for capital that incorporates time value and a risk premium, we use it to discount future cash flows. These are added up to form a net present value.

There are two main difficulties we face in finding the cash flows caused by computer-based systems. The first is their intangibility – the difficulty of translating elements that are unconnected with financial transactions into financial values. And the second is their diffuseness – the complex pattern of causes and effects that links a technology's operational properties with its financial consequences. The best way of tackling both is not to go beserk with algebraic techniques but to leave open the possibility of making simple, human judgements. We can make the most of this subjective process by ensuring that

- it is informed: that decision makers are provided with a comprehensive, qualitative picture of the causes and effects connected with a new development; and that
- it is reduced as far as possible in scope: that the knowledge of quantitative effects is collapsed into a present value before it is weighed in the balance against unquantified effects.

Costs are generally easier to predict than benefits whenever systems or services are acquired through a market. There is then an obvious cash outflow. Where this isn't so – where the services of an internal department are used, perhaps – we have to find an opportunity cost by considering explicitly the alternative uses to which the department's resources could be put.

The effects of computer-based technologies on the two factors that determine risk tend to act in opposite directions. Risk is reduced by the flexibility of new systems, which can at least partly offset revenue sensitivity. New systems help overcome what might be called the inertia of mechanized plant and the inertia of complex, bureaucratic organization. But risk is increased by the relatively high proportion of fixed costs that firms incur when they introduce new systems: very few of the expenses associated with these systems vary with a firm's turnover.

Notes and references

1 For example, Allen, D. E. *Finance. A Theoretical Introduction*, Martin Robertson, Oxford (1983)
2 For example, Weston, J. F. and Brigham, E. F. *Managerial Finance*, Holt, Rinehart & Winston (1979)
3 This view is suggested by Franks, J. R. and Broyles, J. E. *Modern Managerial Finance*, John Wiley, Chichester, p. 6 (1984)
4 A lengthier discussion can be found in Heyne, P. *The Economic Way of Thinking*, Science Research Associates Inc., Chicago, p. 151 (1976)

5 There appears to be no real consensus on exactly how this premium should be measured. A detailed discussion is provided by Brigham, E. F. and Shome, D. K. Estimating the market risk premium. In Derkinderen, F. G. J. and Crum, R. I. (eds.) *Risk, Capital Costs and Project Financing Decisions*, Martinus Nijhoff, Boston (1981)
6 Franks, J. R. and Broyles, J. E. *op. cit.*, p. 123
7 An alternative approach is that of Certainty Equivalents; for a fairly non-technical discussion see Hodder, J. E. and Riggs, H. E. Pitfalls in evaluating risky projects, *Harvard Business Review*. January-February, pp. 128–35 (1985)
8 The *Risk Measurement Service* is published quarterly by LBS Financial Services. It gives a substantial amount of information in addition to company betas.
9 A correction, strictly, needs to be applied to reflect the tax advantage to corporate borrowing. A description of the issues involved in un-gearing beta may be found in Buckley, A. Beta geared and ungeared. In Ivison, S. *et al.* (eds.) *British Readings in Financial Management*, Harper & Row (1986)
10 Myers, S. C. Finance theory and financial strategy. *Interfaces*, **14**, January-February, 126–37 (1984)
11 For example, Marschak, J. Economics of information systems. *Journal of the American Statistical Association*, **66**(333), 192–217 (1971)
12 This suggestion may be found in Franks, J. R. and Broyles, J. E. *op. cit.*, p. 82

6 Adding options for growth

It is easy to overlook the absence of appreciable advance in an industry. Inventions that are not made, like babies that are not born, are rarely missed.

John Kenneth Galbraith *The Affluent Society*

6.1 Present value limits

Assumptions of commitment

The previous chapters have looked at ways of assessing advanced manufacturing technologies in a fairly limited way. They have been based on the idea that once a specific development is initiated, the business is committed to a particular pattern of cash flows – a mixture of benefits (perhaps improved efficiencies) and expenses. There will be a number of ways of going about the development, but only one will finally be chosen, and this will determine the nature of the outcome.

There will, of course, be a degree of uncertainty associated with these cash flows until they are in fact observed. This uncertainty is itself partly incorporated in the evaluation of the development through the discount rate. It reflects a lack of knowledge within a company of various factors – like the prevailing economic climate – over which the company has very little influence. But the evaluation process essentially assumes that, given the state of the world at the time the development is put in train, there is one particular pattern of cash flows that unfolds as a result: even if at that time it is imperfectly understood.

Missing options

The problem with this assumption is that it ignores the fact that when real decisions are made they rarely close off all a firm's options. It is quite important in fact that a certain amount of discretion is left open to the people who work on projects of any scale. Since they don't have perfect foresight, and since it is never feasible to collect all the information that exists before starting work, they will need to meet unforeseen contingencies from time to time. Without an ability to improvize a little they would find this very hard. For developments that are in some way unusual – maybe because they embody an up-to-date technology – it is more expensive to collect information and knowledge that might, for the most part, lie outside the organization. It is therefore reasonable to expect that advanced technologies will spring more than their fair share of surprises, and that they will be applied in a way that leaves good room for manoeuvre. Perhaps the course of a project can be changed during its early months. Perhaps it can be stopped at some point and resumed after a long delay.

We also know that the effects of new systems tend to be profound and wide-ranging, and we know that they are highly flexible. We might therefore expect to be able to develop our systems beyond our original intentions. Perhaps, once a development is complete, a number of new projects can be started which build on the first in some way: or perhaps the development can be modified every so often to track changes in its users' needs. It may be that some types of development narrow down a firm's options more than others. But most will provide opportunities to undertake work that goes beyond the original programme – opportunities that can be exploited (or left unexploited) according to a separate decision made at some point later than the go-ahead given for the initial work. In other words, a project will embody opportunities that can be adopted at a firm's discretion.

Take the example of a local computer network. Suppose that a firm is considering the installation of a LAN in some of its offices in order to replace the direct wiring installed between a number of terminals and the computer running its MRP system. A simple evaluation might consider the benefits of cheaper terminal re-location, cheaper maintenance, and the ability to re-use the same terminal for working at both the MRP system and, say, the process planning system. These benefits would be set against the costs of buying and installing so many metres of cable, so many terminal server boxes, a network management box perhaps, and dropper cables to a number of terminals. The decision about whether or not to proceed would depend on whether the NPV proved to be positive or negative.

But this analysis doesn't incorporate the value of being able to delay the project for a year. It doesn't reveal the value of reversibility – the extent to which the components can be turned back into cash. Some of the boxes might be marketable as second-hand equipment, or they might be useful to other divisions within the company. Perhaps most importantly, the NPV doesn't express the value of having that network in position when the company begins to form more ambitious plans for integration. Without the network it wouldn't easily be able to pass process routes from the process planning system to the MRP system automatically, nor get product bills-of-material from the CAD system. The integration is, if you like, a potentially valuable option that the company can take up once it has a network in position. To say that the worth of the network is equal only to the value that derives from terminal traffic (because that is the extent of the commitment) seems to be missing an important point. The sketch in Figure 6.1 is intended to illustrate this.

Each empty circle represents a specific decision. It is apparent that the first decision (to go ahead with the network) allows a number of further developments to be undertaken. Some of these options, such as liquidation and postponement, are defensive – in the sense that they provide a way of coping with an unfavourable outcome. Others, which build on the network, are much more positive, and demonstrate that it has more worth than simply that of carrying terminal communications.

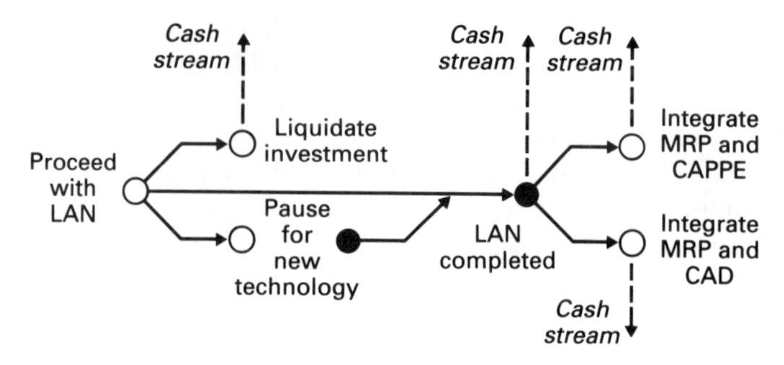

Figure 6.1 Options on a network

The presence of options (other than liquidation and postponement) is plainly contingent on two things:

- first, that there is a certain amount of versatility in the systems that are being developed; and
- second, that there is a readiness on the part of managers and technologists to explore and exploit this versatility.

The foundation of advanced manufacturing systems on computer technology provides a degree of versatility and adaptability that was never available with traditional forms of machinery, nor with the more inflexible forms of automation. Continuing efforts to explore new manufacturing techniques, and to involve everybody in efforts to improve the process of manufacturing, help bring out the will to exercise such options.

Following strategies

A related problem with a mechanical application of the net present value yardstick is that it encourages an analyst to concentrate on the individual investment decision. The appraisal is used to establish whether each proposal that gets to the stage of being seriously considered for adoption is worth taking up. It applies a specific test to the proposal as a distinct entity, and it rules it acceptable or not without any regard to other work that has been started, or that is currently being considered. There is no memory, as it were, in the net present value filter, and it doesn't make anything of the associations between one project and another.

This is not to say that the process of decision making based on a present value yardstick prevents the analyst incorporating his knowledge about these interdependencies. If one project will gain some benefit from being accepted at the same time as another then this benefit may plainly be expressed in the cash flows that are forecast for it. For example, the success of a new FMS is likely to be enhanced if its products have been engineered in a design-for-manufacture project, and the FMS's positive cash flows can be boosted if it is known that this is the case.

But the point is that this identification of mutual cohesion, this idea of a company evolving through a sequence of logically-connected developments, lies outside the net present value structure. The analyst has to remember to look for these connections, and he has to work out a way of converting such imprecise notions to cash flows. This is, in the present value calculation, an entirely unspoken step – a task that must already have been performed by the time the NPV is worked out.

What we really need is an explicit process for evaluating developments that fall within a strategy – the sort of general, far-thinking plan that firms adopt to help keep a multitude of developments focused on a common goal. This process of forming strategies is the subject of a section in Chapter 7, but here we want to see how the evaluation process can take explicit account of any strategy that a firm may already have adopted. Although we could say in the last chapter that the present value yardstick didn't contradict strategic thinking, we could not say that it in any way assisted it, or was assisted by it.

The reason it was suggested that strategic considerations are related to the issue of options discussed in previous paragraphs is that both are concerned with the value that is attributable to a *succession* of decisions. The options issue centres on the need to include in the evaluation some indication of how a development may be built upon in succeeding projects. The worth of a strategy stems from its identification of a sequence of actions that takes a firm in a direction that is consistent in some way from one time to the next.

Take the example of the data network again. A firm might be pursuing an information technology strategy in which it seeks to make as much as possible of its stock of information available to all its information workers. As part of this strategy it may decide to adopt two lesser goals: first, to carry out the work needed to make a physical connection between everybody's terminal and every computer, and, second, to carry out the work needed to ensure that data in all its computer applications is kept mutually consistent (Figure 6.2).

It will, for instance, doubtless want to make sure that product structures in the MRP system are the same as those in the CAD system. When it comes to evaluating the worth of the local network, this again illustrates the need to be concerned about its value as an infrastructure both for terminal access and for integrating applications. It doesn't matter which comes first, (perhaps the terminal traffic which is probably the easier development of the two): it shouldn't just be the first that provides the network's financial justification. The terminal traffic might, of course, be insufficient justification on its own, and if this were the sole basis for the decision on whether to install the network the firm could well find its strategy being unravelled.

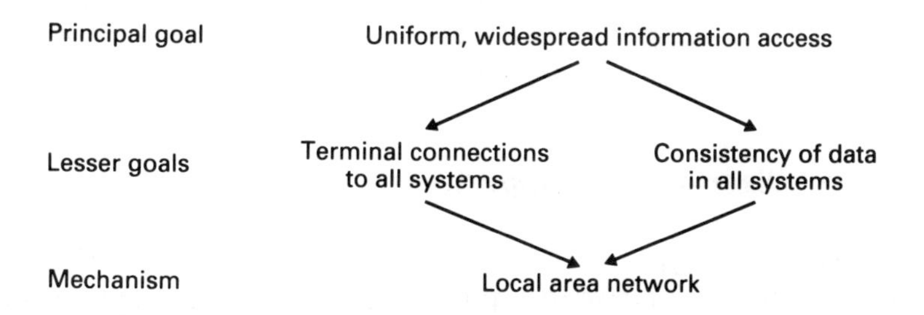

Figure 6.2 Goals and mechanisms

These issues undoubtedly make it harder to reason about the value of a specific project. They interfere with the simplicity and clarity of the net present value yardstick. They re-introduce some of the complexities we might have hoped to eliminate by reducing our financial model to one of simple cash flows and a discount rate. But relevance and usefulness come first, and it would be quite wrong to ignore important issues on the basis that they complicate the appraisal process. In particular, we know that patterns of development are not perfectly decomposable – that one computer-based application is too bound up with another for decisions about each to be made in complete isolation. Options help avoid the notion of decomposability while, at the same time, preserving the distinctions between separate projects that make the process of development manageable.

On the outside looking in

The final arguement for seeking an additional yardstick is the extent to which people outside a company place a value on its assets, as a whole, that exceeds the present value of the company's current earnings. In a paper published in 1984[1] Kester looked at this premium for some large, quoted manufacturing companies in the United States. He first estimated the net present value of the firms' existing earnings by assuming that those earnings would remain unchanged in the future. He then calculated the difference between this net present value and the current market value of each company's equity.

In every case the difference was substantial: the market value outweighed the present value by as much as 88%. Kester's calculations were approximate, but his purpose was to show that the financial markets attributed far more value to most companies than was evident from the NPV of their existing earnings. In other words (barring temporary upsets in profitability) investors were expecting substantial growth. They expected the firms to make far more from some of their existing assets than was currently apparent in their reported profits.

This disparity reflects, in an aggregated measure, the issues described earlier. Having the option to undertake valuable new investments is a source of growth, for growth is simply an increase in a firm's worth. So whatever the assets are that provide these options, they have some intinsic value in addition to the contribution they make to current activities. Even shareholders agree. It is therefore not enough to look at the immediate effects of a new development when you assess whether to proceed with it; it is usually wrong to consider a single pattern of cash flows as its inevitable result. It is important to consider the options that such a development offers a company, and if these options are sources of positive value then they are, by definition, *growth options*.

6.2 Types of discretionary opportunity

There are two main types of option described in this section. The first – the option to liquidate or reverse a project – is applicable to any investment, although there are a few remarks worth making specifically about advanced manufacturing. The second type – the option to indulge in new developments following an initial project – is the most important category, and it covers most of the issues hinted at in Section 6.1. This category is itself divided into two: one part for projects that simply enhance the worth of subsequent developments, and one for those that are essential precursors to further work.

Options to liquidate

This is the pessimistic category. It contains all the opportunities that might allow a company to cut and run when a new development looks like becoming an expensive failure. The extent to which investment in a new development can be liquidated at particular points in its life is clearly something that people do consider during appraisal, if only on a qualitative basis. The essential argument here is that this reversibility can be expressed as an option. We want, basically, to express the fact that a development which offers options to deal with unexpectedly unfavourable outcomes is better than one that doesn't.

Complete *ir*reversibility is the condition in which there is no other use, whatsoever, for the results of the investment. It suggests that there is, in other words, no salvage value: all the equipment, training, software development and so forth cannot be put to use elsewhere. In large part, it is an expression of how specialized those assets are to a particular application. An obvious example is software written specifically for a particular project which, by its nature, is useless for any other project. Systems integration work, for instance, might be largely irreversible because highly specific programs need to be written to provide interfaces between one application and another. Should an application change or become obsolete, then the integrating software (or a part of it) will most likely have to be scrapped.

The value of an option to reverse a project therefore depends on the value that can be salvaged, whether this is appropriated in practice by selling second-hand equipment to other companies, or by re-using knowledge (and perhaps software). The option to reverse will be exercised, in theory, if ever a project's managers come to the conclusion that the net present value of liquidation becomes greater than that of continuing.

The more exposed a project is to uncertainties in events beyond the company's control, the more likely it is that its final outcome will be a long way off the expected outcome: that is, the more likely the net present value in practice will turn out to be much lower or higher than the mean present value forecast during appraisal. Hence the more likely that an option to reverse will be exercised. This means that thinking about the reversibility option is especially important for developments that are highly sensitive to such influences as product demand, or commodity prices. It is also important to projects that have long gestation periods, as uncertainty normally increases the farther in the future people attempt to forecast. Introducing an FMS, for instance, generally takes a good deal of time, and its present value is often subject to bold assumptions about how many units of a specific product the company will sell during the years following its installation. It is also a development whose complexity, and whose novelty, tend to magnify uncertainties.

You might not like to think that questions of reversibility should play a big part in the decision to adopt an FMS. You might feel that recognizing the possibility of an unfavourable outcome lends such an outcome a degree of legitimacy, and reduces the drive to make the work a success. But it is at the very least a factor that enters into the choice between different types of FMS.

In practice, of course, every project develops a good deal of momentum and it is unlikely to be terminated just as soon as its prospective net present value falls below that of liquidation, if it ever does. For an option to be genuinely an option it must plainly be an organizational possibility as well as a technological one. Hard hearts are as much a prerequisite as clear minds.

There are dangers, in any case, in being analytical to the point of overlooking free-will and determination. There is no development whose performance is entirely beyond the control of the company that undertakes it, and it may be that the position of a project whose prospective return is marginal can be recovered by some inspired leadership. It may be that a degree of enthusiasm and imagination can confound a simplistic appraisal. In short, when managers come to review the position of a project, with the intention of deciding whether or not to liquidate it, they are not faced with two, simple figures: the NPV of continuing and the NPV of terminating. They are faced with two, fuzzy bands of likely outcomes – outcomes that are, to a degree, within their influence.

A rather more intricate study of reversibility, postponement and the capital budgeting process may be found in a paper by Stark.[2]

Options to grow

This second category is the more interesting one. It includes all the opportunities for growth that become available once a particular development has been successfully concluded.

Growth, in the broadest sense, simply means some addition to the net worth of a company over a certain period. For example, the adoption of flexible approaches to manufacturing can help a firm look to new markets (perhaps for high-value, low-volume products) which were formerly closed to it. This comes on top of the benefits that are due to the application of flexible manufacturing to an existing product mix – perhaps smaller batches, and the better quality and shorter lead-times that should follow.

It is worth spelling out how these options for growth differ from simple considerations of net present value. Suppose that by undertaking project A a company then had the opportunity to adopt at its discretion project B. The projects are expected to have net present values of $NPV(A)$ and $NPV(B)$. Then the present value investment rule says

Go ahead with A if, and only if, $NPV(A) > 0$:
and go ahead (later) with B if, and only if, $NPV(B) > 0$

What we cannot do in this rule is to attribute some part of the value of B to A, even though A may be a fitting predecessor to B. The reason is that the two are distinct projects, and that there is no commitment to adopting B if A is adopted. What we would in fact like to say is

> *Go ahead with A if, and only if,* $NPV(A)$ + *value of an option on* $B > 0$:
> *and go ahead (later) with B if, and only if,* $NPV(B) > 0$.

This rule captures the fact that A does something more than generate cash directly. It also helps to show how A can be a component of a continuing programme which extends much beyond A. Still further, it prompts us to look at existing investments (developments of type A that are already in place) to see what options for growth they might offer. Such issues are the stuff of strategic thinking, and they capture well the process of continual and partly uncertain development that lies at the heart of most businesses.

There are two distinct types of these options on growth. In the first case, one development attaches, as it were, extra value to a subsequent development. The existence of an engineering data management (or EDM) system might give some additional worth to a design-for-manufacture initiative. Design-for-manufacture is therefore a growth option that may be attached to EDM. The key factor is that while EDM makes design-for-manufacture more attractive, by automating controls over the life-cycle of product designs perhaps, it is unlikely to be an absolute neccessity. It is, in other words, probable that design-for-manufacture has a positive net present value on its own: it would simply have more if it were built on the infrastructure of engineering data management. It is the option on this additional parcel of value that is being acquired. In fact, because one development is complementary to but not contingent on the other, we can reverse the logic and make design-for-manufacture the initial project and EDM the growth option.

The second category, as you might suppose, is that of options which would not be feasible at all without some prior development having been successfully completed. While it has become something of a banality, it is now widely recognized that systems such as engineering data management (and computer-supported co-operative work in general) are valueless without some preceding simplification of administrative procedures. Engineering data management is a prominent option following a simplification project, although it remains just that – an option. But it is probably true to say that it is an option that would not realistically exist without the simplification (Figure 6.3).

6.3 Options arising from technology

It ought, by now, to be apparent that growth options play a role of central

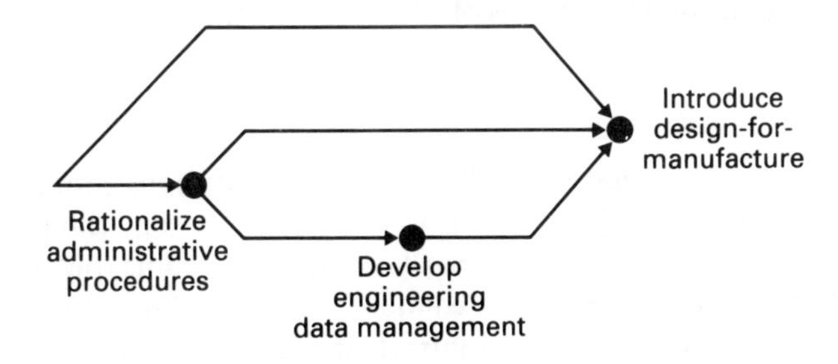

Figure 6.3 Options on engineering developments

importance in the process of understanding the new technologies that are the subject of this book. Considering such options is logically a part of the appraisal of any new investment. But it is in the nature of newer technologies that growth options loom larger than they did with the more fixed types of capital plant. Before tackling the detailed analysis of options, it is worthwhile looking quickly at their typical sources.

Knowledge

An issue that seems to have captured a few important imaginations recently is the contribution that effective education can make to successful business. Whether this is a new preoccupation I find hard to say, but there is little doubt that with more open markets, and advancing technologies, people need a good deal of prior knowledge in order to do their jobs well. It is not enough that they have the knowledge specific to time and place that once might have sufficed. To know only the ins and outs of their company's procedures, organizational structure, people and machines is to know too little. They need to demonstrate the grasp of techniques that enables them to *change* the nature of a company and its operations: whether this is at the detailed level of the shop-floor quality circle, or at the more general level of long-range planning.

The scope of this knowledge is very general. It includes the knowledge of engineering processes, of manufacturing processes, organizational techniques, systems and so forth. It also includes some less tangible things, like an understanding of people and organizations. And it includes some very general elements, like literacy, numeracy and commercial insight.

If a good stock of this knowledge exists, however that might be measured, it provides innumerable opportunities for growth. In particular it is both a requirement, and a product, of any technology that is new, complicated and malleable. Computer-based systems are therefore rarely successful when the people who work with them lack a substantial expertise. And the experience of working with these systems (if it isn't lost in disillusionment) can expand greatly the stock of knowledge applied to subsequent work.

These arguments have become trite. But proposals advanced in support of education are rarely accompanied by a specific financial evaluation – mainly because people don't know precisely how knowledge will be applied over the months or years for which it remains applicable. Again it is the problem of having options, rather than committed lines of development, that makes appraisal difficult.

Infrastructure

The next major class of investments that yield growth options contains what might be called infrastructures. Typically, an infrastructure is something of reasonable longevity and scale. It is something that lacks intrinsic value, but something that serves as a foundation for more immediately useful developments.

A computer network is a piece of infrastructure, and it contributes nothing to a firm's earnings stream unless it is being used for communication (although it might be worth something on the second-hand market). The same applies to the higher-level integration developments, such as distributed databases, that make the dispersion of information to different sites transparent to the people that use it.

Because it has little value in isolation, a project developing a new infrastructure relies almost entirely on growth options for its justification. When an infrastructure is retro-fitted to an existing superstructure (such as the application of a new network to old computers) these options tend to be exercised immediately and automatically. But because people generally take pains to make networks and integration mechanisms highly extensible, there will remain a surplus of growth options which are not taken up straight away.

Flexibility

The flexibility that we have good reason to expect from computer-based technologies is potentially a rich source of growth options. It is in fact a tautology to say so, because flexibility is hardly flexibility if it fails to offer worthwhile options. Flexible manufacturing systems usually allow a factory's managers a certain amount of discretion over both the mix of products they will produce, and over the manner in which they are processed. They will have a range of alternative products and process routes from which they can, in theory, choose the one giving the best combination of revenue and expense.

These are short-term options, in that the decision to exercise them is unlikely to have long-lasting consequences. Much more fundamental in nature are the opportunities that flexible systems offer a firm for changing the nature of its business. The earlier chapter that discussed economies of scope, and markets and hierarchies, looked at these long-term developments in detail. Economies of scope are, by definition, the gains in efficiency that come from applying sharable resources. By buying an FMS or an expert system you are, so to speak, buying an option on these efficiencies. The reason that it is an option, rather than a commitment, is that the economies stem from *scope*. or variety, rather than flexibility in a direct sense. Scope is a function of the products that a firm chooses to produce, while flexibility is a function of the way it chooses to organize its operations. Flexibility is, as it were, a precondition for scope – but it does not guarantee it. It does not have to be exploited at all if it eventually becomes obvious that to introduce greater product variety would not be economically attractive.

Flexibility arises because computer-based systems can embody complex decision-making structures, and because (as we have already seen) these structures are highly malleable. The first of these properties yields mostly short-term options, while the second yields the longer-term options.

6.4 Growth option analysis

This section looks at growth options in a more analytical manner, and ultimately it shows how the value of such effects can be quantified. The reasons for taking this step are the same as those that were used to justify the use of net present value: the need to be able to compare the worth of one development with that of another (and with that of doing nothing); the need to have a repeatable and consistent yardstick; the need to communicate in a common economic language inside and outside the firm. It will become apparent, however, that this step is sometimes a difficult one to take in practice. There will be instances where we shall not want to advance beyond the kind of qualitative understanding reached in earlier sections. Formal models are often useful in making the concepts more precise, but they aren't always suited to practical decision making.

An analogy

The basis for the analysis is the analogy that can be drawn between growth options on industrial investments and call options on financial securities[3]. Since the latter are the subject of large-scale, quantitative analyses we might expect to find analytical methods that we can re-apply to the former. There are bound to be some differences between the two types of option that eventually complicate things, but if the key variables in the two cases behave in similar ways the analogy will be good enough for our purposes.

A call option on a financial security such as a share entitles the option's owner to buy the share at some point (the *maturity* date), or during some period, in the future. He will be able to do so at a price (the *exercise* price) fixed at the time the option is first sold. Since an option is an entitlement to buy an asset, and not a commitment, it will only pay the owner to exercise it if the share price at the date the option matured is greater than the exercise price. The owner will have gained, in the long run, if the share price is sufficiently greater than the exercise price to cover the price he paid for the option, any transaction costs, and any loss in value due to elapsed time.

By analogy, a growth option allows its holder (an industrial company) to acquire a profitable asset at some point in the future at a specific cost. The final point at which the asset in its envisaged form can be acquired is the maturity date, and the cost of acquiring it is the exercise price. The benefits that can be obtained from the industrial asset when it is acquired are analogous to the price (that is, the value) of a share at the time a share option is exercised. These benefits are summarized by the net present value of the cash stream due to the asset *after* it has been acquired.

Taking the example of an earlier section, suppose that a firm is assessing the value of installing a data network, with the eventual intention of integrating applications like CAD, CAPP and MRP. Then by introducing the network it is taking out an option to acquire the associated integration mechanisms. The integration work might need an extra investment of £100000 (once the network is installed), and it might be expected to generate a subsequent cash stream having an NPV of £200000. This option might mature in, say, three year's time when the applications are likely to be obsolescent. The option to integrate that comes with the network therefore has a maturity date three years hence, an exercise price of £100000 and (on current predictions) an underlying asset value of £200000.

Option value principles

We now need to consider how to go about calculating the value of an option that enables such potentially profitable assets to be acquired. In the example just

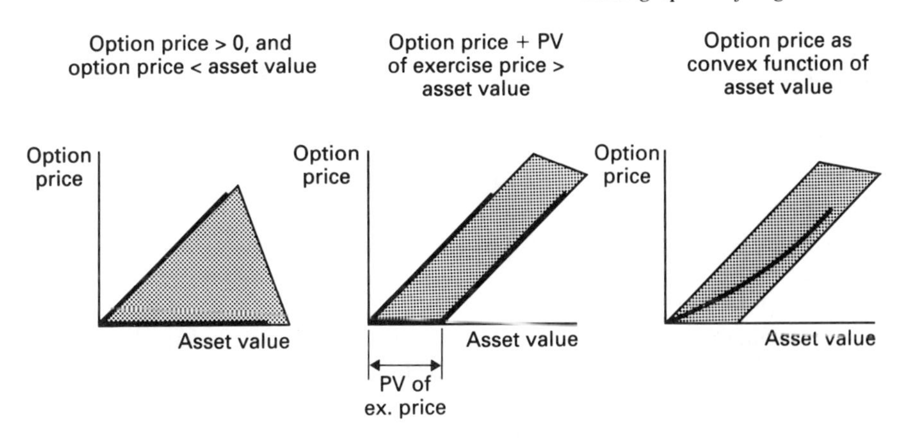

Figure 6.4 Option price, asset value and exercise price

outlined, we would want to know how much it is worth to us to have the opportunity of investing £100000 in integration work. One of the advantages of looking at options on *shares* is that the values of the assets that underlie them – that is, the shares themselves – are readily observable in a market. Another advantage is that options are traded in their own right: so we can reason about option prices by thinking about their values in a competitive market.

The ideas described in the remainder of this sub-section are well-known and, as usual, they can be found in a much more complete and rigorous form in finance textbooks[4].

The first thing to do is to see how the value of an option depends on the value of the underlying asset and on the exercise price. And, to begin with, we can predict that the option's value will never be negative since it does not commit its owner to doing anything – it is only an entitlement to acquire an asset. Someone wouldn't be paid to accept an *absence* of obligation.

We can also say that the option is never priced more highly than the asset that underlies it: for if it were, it would always be cheaper to buy the asset itself and the option would never be traded. You would, hopefully, never pay £200 to have the option of acquiring something worth £100. It would be better to buy the asset and throw it away if it subsequently turned out to be valueless.

A third property is that the option, together with the present value of the exercise price, is at any time more valuable than the underlying asset. This is a more complicated rule, but it can be understood by bearing in mind that the present value of the exercise price is the amount of money that, if we had it now, would allow us to exercise the option at its maturity date. The combination of option and present value of the exercise price would allow its owner to buy a share (so that he can take advantage of any appreciation in its price). But it clearly relieves him of the loss if its price falls. It is simply better to buy the combination than to buy the asset itself.

These three factors set a boundary to the form that the relationship between option price and asset value can take. Together with one or two additional theorems, the relationship can be shown to look like the one indicated in Figure 6.4. The first sketch illustrates the effects of the first two rules, allowing the option price for a given asset value to lie only within the shaded sector (between the horizontal axis and a 45 degree line through the origin). The second sketch shows the effect of the third rule which narrows down the relationship to a band that points towards the top right-hand corner of the plane. In the third sketch, some principles that have not been described give the relationship the form of the curve drawn there.

This quick analysis has so far ignored transaction costs. These are the costs associated both with collecting information and with making exchanges. For example, a potential buyer might spend some time and go to some trouble to find a person or institution who will sell him an option that exactly suits his purposes, and the seller (and any broker) will want to charge him a part of the administrative costs that they incur. In the case of the industrial investment, there may be a substantial cost for *searching* (that is, for identifying possible investment opportunities), and there will be costs for *evaluating* such opportunities.

Figure 6.4 suggests that the lower the present value of the exercise price, the more valuable the option (other things being equal). This can be seen in the second and third sketches: as the present value of the exercise price falls, the right-hand boundary moves towards the left, pushing the curve upwards for finite values of asset price. In a growth option, the exercise price has an analogy in the form of the capital investment that has to be made to initiate the optional development. Suppose this is I, and suppose that it is known in advance with certainty, and is therefore risk-less. Then if the risk-free rate of interest is r, the present value of the investment in n years' time is $I/(1 + r)^n$. This expression demonstrates that the lower the exercise price I, or the greater the number of years to maturity n, or the higher the risk-free rate of interest r, then the lower the exercise price present value and the more valuable the option. However, since the option value cannot exceed the price of the underlying asset, the value doesn't climb indefinitely as either n or r increases: it can only approach the asset price. (These arguments are complicated by the fact that some options can be exercised over a period of time rather than at a specific maturity date. Fortunately it is the case that such an option has the same value as an option that can just be exercised on a single date. *Un*fortunately this is only true, in the case of share options, when the underlying share doesn't pay dividends: when it does, the option that can be exercised over a finite period is more valuable. But, for the purposes of our analogy, this difficulty will be ignored.)

The last parameter on which an option value depends is the uncertainty in the final value of the underlying asset. If this uncertainty didn't exist at all an option would be valueless since one could make a definitive choice of actions at the outset. In the case of share options, this uncertainty is a result of the lack of perfect knowledge in the present about the share price at the future time at which the option can be exercised. The nature of the relationship is such that the greater the uncertainty about the value of the investment, the *more* valuable the option. (Again, this is subject to the qualification that the option value can only approach the underlying asset price: it cannot exceed it.) The form of this relationship is a little surprising given the risk-return trade-off mentioned in the last chapter, but it stems from the fact that an option is just that – optional. The more uncertain the expected asset value, the greater the likelihood that it will be either much higher or much

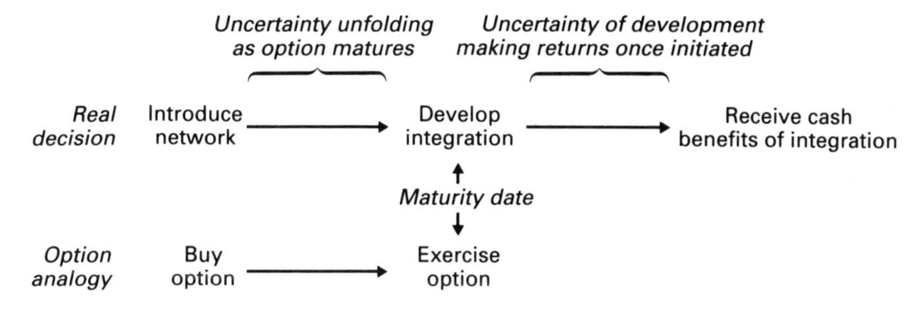

Figure 6.5 Uncertainty and option value

lower than the expected value. If the asset value declines between the present and the maturity date, the option can simply be left to lapse, and the risk on the down-side can be ignored, irrespective of its scale. Any gains, however, can be taken up by exercising the option. The more uncertainty there is, the greater the likelihood of large gains.

It is important to understand that this uncertainty is the lack of knowledge that unfolds between the time the option is acquired and the time it is exercised. In the case of an industrial investment, the exercising of an option represents the initiation of a new development which still embodies a degree of uncertainty, even at the time it is started. This remaining uncertainty does *not* have the same effect on the option value as the uncertainty associated with the maturing of the option.

For instance, to continue with the earlier example, suppose that somebody makes a rough estimate that the expected value of integration, once a network is installed, is subject to a variance of maybe 40%. That is the extent of the uncertainty between now and whenever the integration work is done (it might be a year or so away at the moment). It may, in addition, be decided that even when the integration work has been completed there will still be an uncertainty with a variance of 25% associated with the forecast of its subsequent benefits. Here, it is only the 40% that has a positive effect on the value of the option to integrate. The 25%, on the other hand, doesn't directly affect the option value at all. Figure 6.5 illustrates this idea.

Advanced technology and the input parameters

The previous sub-section identified, if only in qualitative terms, the parameters that set the value of an option. First, the option value increases with the value of the asset that underlies it. Second, it falls with the price that must be paid to exercise it. Third, it increases with the rate of interest, and with the time available before it matures and must either be exercised or allowed to lapse. And, fourth, the option value rises with increasing uncertainty about the asset value during the period leading up to the time when the asset is acquired. By seeing how computer-based technologies affect each of these parameters, we can get some idea of the worth of any associated growth options.

The fact that the value of an option increases with the prospective asset's value, and that it decreases with the exercise price, simply suggests that developments with the greatest net worth have the greatest option value (other things being equal). Options on integration work, for instance, will be most valuable where there is a large volume of shared information between different applications, where there is a high probability of inconsistency between replicas, or where there is high cost associated with this inconsistency. Volumes are likely to be highest where there is a duplication of *function* between different applications. For example, the storage of product structures is commonly found in CAD-CAM repositories, process planning systems and MRP engineering databases. The costs of inconsistency might be highest in areas where the largest proportion of manufacturing expenses are determined – typically during product and process engineering activities. The probability of inconsistency is perhaps highest where updates are frequent but transcription irregular.

A part of the value of this type of development reflects the extent to which it gives a company a unique advantage over competitors. If, in other words, a development has some highly distinctive effect other than generally reducing a firm's cost base or improving its delivery performance, it will be especially valuable. Kester[5] distinguishes options on such assets by calling them *proprietary*. Options which don't demonstrate this uniqueness are *shared* in the sense that competing firms have just the same ability to acquire such options. This difference is perhaps less important for process developments than it is for product innovation. Nonetheless, a major determinant of the extent to which opportunities are shared is the speed at which know-how diffuses from one firm to another. There is an argument[6] that launching a new product entails a more explicit disclosure of knowledge than the adoption of a new prodution technology, so it may in fact be the case that production technologies offer more proprietary options than product technologies.

As for the period to maturity, it is less easy to make generalizations. Some options will be available, and the value of the underlying asset will be undiminished, for a good period of time. But it is worth considering whether these options are contingent on developments other than the obvious precursor. For example, the option to build an FMS following a group technology project might rest on the assumption that a particular product range is to continue in production. It may cease to be a valuable option some time before the product range is discontinued. In areas where technologies are changing very quickly – in personal computing and networking, say – the period over which options are available may be set by the speed at which systems become outdated as a result of their inability to communicate with others.

Uncertainty is another parameter about which it is difficult to say anything very general. Typically, the sorts of opportunity that embody the greatest uncertainty are those of great scale, those involving novel technology or technique, and those that are associated with radically new organization and working practices. Systems such as CAD-CAM and FMS, whose benefits are manifest in revenues rather than costs, and whose technology is often unprecedented in the firms that adopt them, are characterized by the uncertainty in their returns. As far as options on their introduction are concerned, this uncertainty is now a contributor to worth: it is a source of opportunity to a greater degree than it is a source of anxiety. But one has to remember that the uncertainty relevant to option valuation is only the lack of knowledge that will be resolved before the point at which the development is done. Thereafter, the remaining uncertainty can only reduce the development's value.

An option valuation model

At this point it is worth looking at a standard formula for valuing growth options. If you think that the application of arithmetic to decision making is inappropriate you might want to skip this and the following sub-sections.

The formula is, in its appearance, fairly simple, but its derivation is far from straightforward; I shan't attempt to give any justification for its form here, and details can of course be found in the literature[7]. It is based on arguments about markets in financial options, and in carrying it across to the industrial case we lose something of its applicability. It illustrates, nonetheless, the way that we can reason about optional developments in figures rather than words. Some of the difficulties and practicalities that attend its industrial application are discussed in the next section.

Assume that a growth opportunity is the option to invest an amount I at time T to create an asset of value $A(T)$.[8] T is obviously the date at which the option can be exercised – the maturity date. I is assumed to be fixed in advance, but the value of A varies with time. The investment I is analogous to an exercise price, and A to the value of an underlying security. When the option expires, at T, the rational decision is to exercise the option by making the investment I if the asset value at the time exceeds the investment, that is, if $A(T) > I$.

Strictly speaking, there are a number of assumptions that need to hold if this formula is to be valid. Whether these are warranted is discussed briefly in Section 6.5.

The formula can be applied to obtain the value of the growth opportunity at any time t that precedes the maturity date T. This value, G, is a function of the expected asset value as it is perceived at time t, $A(t)$, and of the investment that will be needed at the maturity date I:

$$G(t) = A(t)N(d_1) - I\,e^{-r\tau}N(d_2)$$

where $\tau = T - t$, $N(d_i)$ is the area to the left of d_i under a normal distribution of unit variance and zero mean, and

$$d_1 = \{\ln\,(A(t)/I) + (r + 0.5\sigma^2)\tau\}/\sigma\tau^{0.5}$$

$$d_2 = d_1 - \sigma\tau^{0.5}$$

Here r the risk-free interest rate and σ^2 is the variance of the anticipated return on A.

Although the statistical terms in the expression for G, of the form $N(\bullet)$, are a little opaque, you can see that without them G is simply $A(t) - I\,e^{-r\tau}$. This is just the difference between the value of the asset expected at time t, $A(t)$, and the present value of the exercise price, $I\,e^{-r\tau}$, expressed in continuous time rather than discrete years. This reflects the qualitative principles discussed in an earlier sub-section. The statistical terms are introduced to account for the (supposedly random) process that connects the current asset value with the uncertain asset value at the maturity date. The value σ^2 represents the variance of this process – the breadth of uncertainty, if you like, in the expected asset value at a point in the future. It is often found, in the case of options on shares, by looking at the past fluctuations in the value of shares, and by assuming that this will be a good guide to their behaviour in the future.

An example

The example that follows is a contrived one because it is intended just to illustrate the application of the option valuation model. A real instance will involve some far less clear-cut reasoning, and probably considerable difficulties in estimating values for some of the quantities needed here. The distance between principle and practice is again an issue that is addressed in the closing section.

Suppose that a firm has installed a computer network, and wants now to develop a scheme for linking its main office-based applications – notably CAD-CAM, CAPP, MRP and some shop-floor cells. It might, for example, have introduced some distributed database mechanisms which ensure that replicas of any piece of information in any of the systems are always updated together, or not at all. In any case the guts of the technology are unimportant here.

The appraisal of this development will certainly include an estimate of the present value of the benefits that come from retro-fitting the new mechanisms to existing applications. Maybe these include a reduction in the labour associated with manually transcribing data from one application to another. No doubt they will also include the reduction in rework that may be expected by eliminating errors in manual transcription.

However, on the basis of what has been said in earlier sections, to take the evaluation as far as this and then stop would be to take too narrow a view of what the new scheme is worth. An attempt should also be made to consider the options that the installation of the scheme would offer. Any subsequent development that gains from, or relies on, this integration work would be such an option. Just as integration is an option that comes with a network, so we can expect there to be options that come with integration.

One possibility is the introduction of a product database – a system which, we might say for the sake of argument, has little value if it lacks a close integration with other applications. Its part geometry must be accessible to the company's CAD systems, of course; its CAM-generated part programs to DNC systems; its product structures to the MRP system and so forth. The main aspects of a product database that make it an obvious growth option are its dependence on the integration infrastructure, its scale, the fact that it has a substantial exercise period (a few years perhaps), and the fact that its expected returns are far from certain.

We can now see how we might attach some figures to these quantities (bearing in mind that any apparent tidiness in the figures may not reflect a tidiness in the underlying reality). The benefits the system would have as a result of

(1) quick and simple access to documentation,

(2) the increase in re-use of engineering designs, and
(3) the improvement in management information,
might be associated with an expected present value of £0.5M. (This is the present value of the cash flows *except* for the initial capital investment.) This could be contingent on the investment being made in the next two years, a time beyond which it is likely to be impracticable due to the obsolescence of the integration technology compared with that of the product database.

It is very difficult to estimate variances subjectively with any confidence, but it helps to remember that there is a two-thirds probability that outcomes fall within a range of one standard deviation from the expected value. (The standard deviation is the quantity σ in the valuation formula.) We might, in this instance, assume that there is a two-thirds probability that the anticipated asset value will fluctuate within a range of 30% of its expected value of £0.5M.

Suppose also that the risk-free rate of return is 10%, and that the fixed value of the investment needed is £0.4M. From the previous sub-section, the value as a growth option is then

$$G(t) = A(t)N(d_1) - I\ e^{r\tau}N(d_2)$$

Take time t to be the present, and

$A(t) =$ £0.5M (the currently expected value of the asset underlying the option),

$I =$ £0.4M (the investment that will be needed to exercise the option),

$r = 0.1$ (or 10%, the risk-free rate of return),

$\tau = 2$ years (the time between now and the maturity date),

$\sigma = 0.3$ (the standard deviation of movements in the expected asset value before maturity).

So

$$G = 0.5\ N(d_1) - 0.4\ e^{-(2 \times 0.1)}\ N(d_2)$$

in units of £M. Now

$$d_1 = \{\ln(A(t)/I) + (r + 0.5\sigma^2)\tau\}/\sigma\tau^{0.5}$$

$$d_2 = d_1 - \sigma\tau^{0.5}$$

So substituting the values suggested,

$$d_1 = \{\ln(0.5/0.4) + (0.1 + 0.5 \times 0.3 \times 0.3) \times 2\}/(0.3 \times 2^{0.5}) = 1.21$$

$$d_2 = d_1 - (0.3 \times 2^{0.5}) = 0.79$$

From tables of the normal distribution,

$$N(1.21) = 0.89 \quad \text{and} \quad N(0.79) = 0.79$$

So

$$G = 0.5 \times 0.89 - 0.4\,\mathrm{e}^{-(2 \times 0.1)} \times 0.79$$
$$= £0.19\mathrm{M}$$

Provided that it was correct to assume that the product database was an opportunity that arose from integration, the value of the option, £0.19M, *is* attributable to integration. In reality, it is probably pressing things too far to say that a product database is an option that only exists because of integration: it might be perfectly feasible to move information in and out of the product database manually. If, however, a product database with manual interfaces is worth less than one to which automated transfers are available, then there is at least an option on the *extra* benefit that is bought with the integration infrastructure.

6.5 Practicalities and difficulties

Finding opportunities

Now that the process of evaluating a development is to consider options as well as committed cash flows and risks, the *search* problem becomes lengthier and more costly. Not only do managers and technologists have to cast around for single projects and ways of implementing them, but they also have to identify the options that come with each of these possibilities. So a good question to ask is whether there is any way of economizing on the effort needed in this search process.

One approach is to look at the distinguishing capabilities of a technology, and to examine whether in combination these could be used to provide a new kind of facility. For instance, a firm might be considering the options that are available following the introduction of an MRP system. By noticing that MRP records information about product sales, that it contains a good deal of purchasing data, and that it aggregates and schedules resource requirements, one might come up with the idea of using it to predict a factory's cash flows. This would at least render some consistency between cash flow predictions and the manufacturing workload. Ideas for locating latent possibilities in this fashion can often be found in books about product innovation[9].

Another approach is simply to have a fixed list of headings, against each of which one looks for new applications of a technology. For instance, a recent article in *The Economist* newspaper[10] looked at some general ways in which companies can behave in a creative fashion, instead of simply reacting to the behaviour of competitors. It suggested five kinds of strategy, each concerned with a firm's processes rather than its products: encouraging experimentation, turning specialists into generalists, breaking down hierarchies, unsticking information, and making time to think. Similar headings might be used for the options search. A new process planning system could, for example, help level out the organization of the production engineering department by making process know-how available to everyone within it. It might help unstick information by telling managers how often process routes are designed from scratch and how many work-centres are used only for obsolescent products. And it might make time for production engineers to think by automating some of the clerical work associated with detailing a process route. These particular headings may not quite hit the mark in all firms, but the principle is there.

Calculating growth options and quantifying their parameters is a harder thing altogether, and the next sub-section briefly considers the difficulties.

Problems of application

The correctness of the option pricing formula depends on a number of assumptions. These mainly refer to the nature of the markets in which options and their underlying assets are traded, and they also concern the random processes by which the assets change in value over time. Since growth options are not traded in the same way as financial options, it is hard to carry across the reasoning that underlies the formula and maintain any degree of rigour. Even the qualitative arguments that preceded the introduction of the formula in the last section were based on describing the conditions in which people would buy and sell options in well-informed markets.

The most we can really say is that *if* growth options were traded between firms then our valuation formula would be a good model for predicting the prices they would fetch. A price is as objective a statement of an option's value as you are likely to get. The problem is a little like that of valuing a family heirloom that isn't for sale. You might work out its value by comparing it with goods you know to be traded on open markets – but if you never attempt to sell the heirloom you can in fact attach any value you like to it.

There are, in any case, some practical difficulties associated with the model's application. The first is that growth options do not have precisely-known exercise dates.[11] In the case of share options, the exercise date *is* precise, because it is within the control of the person writing the option. The period over which a growth option is available is determined by factors normally beyond the control of the company that can take advantage of it, and it is rarely very predictable. A growth option might not in fact have a definite maturity date at all: it is more likely to be something whose value diminishes over a period of time. The opportunity that underlies it could lose a portion of its value after, say, a year, then a bit more after something like 18 months to two years, and so forth. The model of the previous section asks uncompromisingly for one date, however, and without a much more elaborate formula we cannot handle gradual obsolescence.

The second problem is that of quantifying the parameters of the option. This problem is one of inadequate information rather than an inadequate structure into which it should slot. But it is at least as big a problem because the value of the growth option that emerges from the quantitative model is entirely dependent on these parameters. The formula asks, for instance, for the present value of an optional development that is at least two decisions away. This will be a good deal harder to provide than the NPV of the investment that confers the option. Perhaps still more difficult is the process of estimating a variance for the uncertainty faced between the time of evaluation and the time at which the option matures.

A third problem is that in reality the costs of a devlopment aren't known in advance with perfect certainty. The formula we have looked at assumes that there is a fixed exercise price. The modification described by dos Santos[12] allows one to cater for this explicitly, but it contains two variances (one for benefits and one for costs) and a covariance. Given the difficulties we already envisage with finding a single variance, this is discouraging. By admitting real conditions we make the analysis more complicated with very little compensation.

The predictions made during a growth options analysis are also hard to test in retrospect. This means that if an analyst consistently predicts a variance of 50% there is no way of knowing that he is approximately right, even after the maturity date. Without this feedback, he will be unable to calibrate his judgement: he will have no way of adjusting it to remove inaccuracies and misapprehensions.

Informal uses

In many instances there will probably be too many difficulties associated with the growth options calculation to make it worthwhile applying. This does not, however, disqualify growth options from qualitative applications. The concepts on which they are based are precise enough, and a good deal more precise than simply claiming that an investment is strategically desirable. The fact that planning for the future is inherently uncertain suggests that there is all the more reason to use a language that gives the issues a clearer, not a fuzzier, definition. Casting the argument for strategic benefits in terms of growth options not only makes the argument more compelling, but it may also reveal options that were previously obscured.

In the same way that an analyst is expected to provide a schedule of cash flows when carrying out an appraisal, he can list a development's growth options. Against each option he can describe, in words if not in figures:

- the value of the underlying asset;
- the exercise price;
- the likely period of maturity; and
- the extent of the uncertainty that will be resolved when the option matures.

In our earlier example of the options associated with integration mechanisms we looked at a product database. Rather than quantifying these variables we might, to be practical, just have described the factors that influenced them. For instance, the period of maturity depends on technological obsolescence – on the length of time for which the integration technologies remain compatible with the applications they are intended to link.

In principle, the valuation formula can be used for making comparisons without going to the trouble of performing calculations. If we could point to the fact that of two alternative developments one had the more valuable growth options then we have a basis on which to choose between them. One of the developments might offer an option that lasted for considerably longer than the other before maturing, perhaps. (This is the value of adopting up-and-coming rather than traditional operating systems: they may or may not offer more services, but they almost certainly allow one to take up more opportunities in the future.) If all other things were equal, this greater maturity period would mean that the development had a greater option value. This process of comparison only works, of course, when one alternative dominates the other – when it is obvious that all its parameters are at least as favourable as those of the other alternatives. But, in a similar way to the process described in the previous chapter, it calls only for the use of ordinal scales, and its need for information is therefore much less onerous than that of a full calculation.

There is little question that growth options exist: they are inherent in the nature of much of advanced technology and in the will of people to steadily improve the processes taking place within their firms. It might prove too difficult to measure them, but it is not too hard to talk about them, to plan for them, and to exploit them.

6.6 Summary

The net present value yardstick is limited to valuing a stream of cash flows in which all are irrevocably connected with one another. This makes no allowance for the optional developments which accompany most investments in computer-based systems: it doesn't let the analyst incorporate explicitly the idea that each development creates a number of new opportunities.

Because a sequence of optional projects defines a pattern of consistent developments, it commonly underlies a strategy of some kind. An analyst can express the strength of the inter-connectedness, and perhaps the cohesion exhibited by the strategy, by predicting the option value added to each project by its successors. Equally, a firm might set out to form a strategy in order to obtain options – perhaps as a way of coping with its uncertainty about future circumstances. Again, the success of this process can be expressed by the values of the options attached to the chosen course of action.

Computer-based systems tend to be rich in both short-term and long-term options. The short-term kind is associated with giving the firm a certain amount of flexibility in the way it conducts its operations – perhaps by offering alternative process routes. Computers ought to offer this flexibility on the basis that they can be programmed to incorporate a large number of decision-making points and, therefore, a large number of possible behaviours. Options of the long-term kind are concerned with giving the firm flexibility in the way it changes its operations – perhaps by speeding up the introduction of new products. The malleability of computer programs, and of information stored in an electronic form, both offer this flexibility. Still longer-term options stem from the way in which computer and communication systems can be linked to each other. This makes it relatively straightforward to build small systems into large ones, perhaps by using networks and distributed databases.

The analysis of growth options is based on carrying across the reasoning applied to options on financial securities. This demonstrates some fairly obvious relationships, such as the fact that the value of the option

- increases with the value of the asset that it lets the firm acquire, and
- decreases with the cost of exercising it.

It also reveals some less obvious relationships. For instance, the value of the option

- increases with increasing uncertainty about the behaviour of the asset before maturity, and
- increases with the length of time before maturity.

Unfortunately, these parameters are difficult to estimate with any credibility, and this limits the usefulness of quantitative formulae. Industrial options also tend to be defined in a slightly different way from financial options, which reduces the applicability of the assumptions on which the formulae are based. Nevertheless, growth options offer us a way of understanding, if not quantifying, an important effect commonly associated with advanced technology.

Notes and references

1 Kester, W. C. Today's options for tomorrow's growth. *Harvard Business Review*, March-April 153–60 (1984)
2 Stark, A. W. Irreversibility and the capital budgeting process. *Management Accounting Research* **1**, 167–80 (1990)
3 This analogy is suggested by Myers, S.C. Determinants of corporate borrowing. *Journal of Financial Economics*, **5**, 147–75 (1977)
4 For example Higson, C. J. *Business Finance*, Butterworths, ch. 11 (1986)
5 Kester, W. C. *op. cit.*
6 Heertje, A. *Economics and Technical Change*, Weidenfeld & Nicolson, London, p. 213 (1977)
7 See Black, F. and Scholes, M. The pricing of options and corporate liabilities. *Journal of Political Economy*, **81**(3), 637–59 (1973)
8 This notation is a simplified version of that used by Broyles, J. E. and Cooper, I. A. Growth opportunities and real investment decisions. In Derkinderen, F. G. J. and Crum, R. L. *Risk, Capital Costs and Project Financing Decisions*, Martinus Nijhoff, Boston, pp. 107–18 (1981)
9 For example Twiss, B. C. *Managing Technological Innovation*, Longman. p.76 (1980)
10 *The Economist*, Management Focus. Create and Survive. 1st December, p.107 (1990)
11 Srinivasan, V. and Millen, R. A. Evaluating flexible manufacturing systems as a strategic investment. *Proc. 2nd ORSA/TIMS Conference on Flexible Manufacturing Systems*, Elsevier Science Publishers, pp. 84–93 (1986)
12 dos Santos, B. L. Justifying investments in new information technologies. *Journal of Management Information Systems*, **7**(4), 71–90 (1991)

Part III The Practice of Investing in New Systems

7 The process of industrial investment

We have to understand the world can only be grasped by action, not by contemplation. The hand is more important than the eye... The hand is the cutting edge of the mind.

Jacob Bronowski *The Ascent of Man*

The arguments in this chapter are much less about issues that can be defined precisely in financial terms, and more about those that influence the way people participate in the process of deciding about new technologies. The language is less unambiguous than that used earlier, and there are few truly general principles on which to build. Much of what constitutes a good investment process depends on the particular circumstances surrounding it. Some organizations, for instance, seem to work better when a single authority issues detailed and unequivocal instructions; others find the act of obtaining a consensus among a wide number of people important in winning commitment and avoiding major mistakes.

In any case, the intention here is to describe something of the sequence of activities in which proposals for introducing new, computer-based systems are taken from their earliest stages to the point where they are firmly adopted or firmly rejected. It is this sequence that the appraisal yardsticks described in earlier chapters are mostly meant to support. Following this there is a discussion of how the adoption of a strategy can lend a degree of consistency and economy to the investment sequence. Finally, the last section examines whether it is reasonable to think that the employees of a firm can really pursue a financial goal single-mindedly, and whether it is worth forming strategies aimed at such a goal.

7.1 The appraisal sequence

The sequence as a whole

Every firm has its own approach to the organization of the appraisal process. Some of the steps are commonly formalized as working instructions, but others are less overt and run on more informal lines. There are several views of what takes place during this process and the purpose of this section is to describe a few of these. It is not always clear whether these views reflect an understanding of what happens in practice, or whether they reflect some kind of ideal towards which a firm is supposed to aim. I have described them in order to examine some of the interesting issues that are likely to need resolving, rather than to suggest that any firm must slavishly follow a predetermined sequence of stages. One needs to maintain a sense of realism about how long a particular organization needs to contemplate an idea before its people are comfortable enough to make it a realistic proposition. And one obviously needs to understand how to capture hearts as well as minds. A known sequence of stages provides a guide to the resources that might be devoted to this process but it doesn't stand in as a substitute for local judgement.

The sequence typically starts with the identification of an opportunity or a problem and the motivation to do something about it. It is a stage you might call *triggering*[1] or opportunity recognition,[2] although the latter suggests that recognition is enough to initiate action – a far from universal case. The identification of possible developments might be a consequence of a systematic search process, or, perhaps more typically, it might happen as a result of a problem becoming so acute that people can no longer ignore it. The disadvantage of a systematic search process is that it is costly and difficult: it demands a considerable degree of acuity on the part of those conducting it, and it may leave them open to a certain amount of abuse for being ahead of their time. A few more words on the search process are given in the next sub-section. The disadvantage of conducting the triggering stage on the basis of acute problems is that there is often every reason to think that it is much too late to address problems only when their symptoms have become acute. With a bit of foresight, and by tackling difficulties early, a company is generally able to minimize the cost of new systems – and to minimize the accumulating losses in its earnings.

When the subject of the appraisal is an advanced manufacturing technology, it is sometimes suggested that more emphasis should be given to the context in which a new development takes place. The idea is that the first appraisal stage should be one in which a company's principal strategies are reviewed,[3] and its current performance assessed. Possibilities for new developments are supposed to become evident in this way.

The following stage is one in which the tentative proposals that emerge from the triggering stage are *screened*. In other words, a certain amount of filtering is done to remove the less promising ideas before substantial resources are devoted to evaluating their effects more fully. In most organizations screening is doubtless an informal activity. It is, however, an important one because a large number of significant decisions are made here, ordinarily on the basis of very little information. In removing a large number of potentially valuable proposals from further consideration, the effects of this process on the future development of the firm are as significant as those of the more formal appraisal that comes later.

As screening remains a rather tacit and localized activity in most organizations, there is a risk that it will be carried out by people who lack an adequate grasp of what is worthwhile to the firm as a whole. There may be a possibility, for example, that opportunities and difficulties identified by people working in production and engineering departments will be shelved by their junior managers. Before screening takes place, new ideas have very little visibility and they are easily suppressed.

The screening process also works with criteria that are never held up to scrutiny, so its development may lag behind that of the formal appraisal procedures. As with budgets, screening is a practical way of dealing with uncertainty and the costs of collecting information. But since it is a matter of practice rather than principle, there is no reason why a particular way of doing it should be very enduring. It is therefore worth re-examining the process (or examining it for the first time) when the circumstances in which it is applied are changing – when it is being used to test new technologies, for instance. This is simply a question of asking junior managers what criteria they apply when they decide whether to allow an investment proposal to be submitted.

The next stage is that of *definition*. Here, a proposal is given a clearer form: its scope is determined, the source of any accompanying economic rent is identified, and perhaps alternative courses of action are listed. Following definition comes a stage of *evaluation*, in which the financial analysis proper is carried out. Although the earlier parts of this book have concentrated on describing a consistent and complete yardstick, in practice it is normal to go to some trouble to present several distinct views of a proposed investment. In principle, net present value plus growth option value encapsulates all that needs to be said about financial worth, but those who deliver the final sanction might also want to know about the payback period. From a theoretical standpoint such a quantity is entirely superfluous and perhaps positively misleading, as we have seen in Chapter 4. But making a compelling case for a new and different piece of work is rarely as simple as making a strictly logical case.

At about this stage, or perhaps earlier, there is a process of *transmission* in which the case for an investment ascends the firm's hierarchy. Early on, this will be relatively informal: a process in which sponsors can be canvassed, and the proposal tested for its alignment with current policies. Subsequently, opinions are committed to paper, transmission becomes more explicit and there is more to be lost if things go awry. Finally, there is a stage at which a *decision* is officially made. By this time, it is perhaps more a confirmation of the results of earlier processes than anything else.

The sketch in Figure 7.1 is a summary of the entire sequence.

The search process

Many developments, including some that substantially change the way a firm does business, are triggered in response to an obvious problem. Some are more a case of exploiting new opportunities than correcting obvious deficiencies. But in many instances a company will blunder on them, rather than detect them in the course of a systematic process. A few projects are perhaps started as a result of technology suppliers pushing their wares. Much as this kind of spontaneous triggering might be valuable, it makes sense at the same time to maintain a continuous and deliberate process of search for new ways of developing the firm's processes and systems. Reacting to evident difficulties is not the most satisfying or necessarily the most profitable way of making progress.

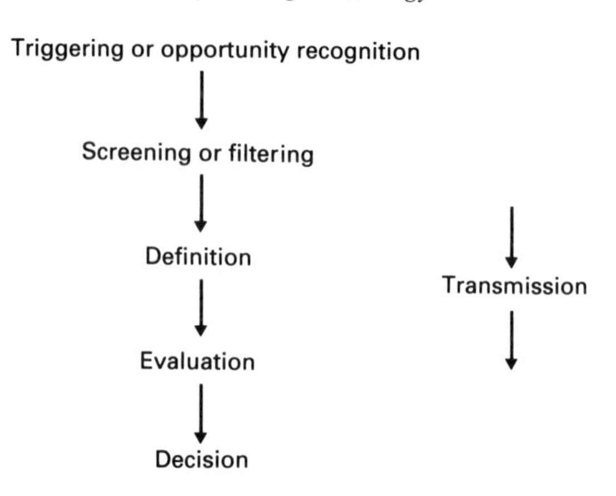

Figure 7.1 A typical appraisal sequence

This search process is, of course, a response to a firm's lack of perfect knowledge about itself and the world around it. Its purpose is to gather information in the expectation that previously unknown opportunities will be revealed. Since the search process itself consumes resources, it is only sensible to undertake it to the extent that these revealed opportunities are expected to have benefits that more than exceed the opportunity costs of the search. Where this point lies is a matter of judgement, not of calculation. But to make the best use of the resources available the process of identifying new investments needs to be guided in some way. This guidance, or framework, is one of the main reasons why a firm should have large-scale plans for the development of its systems. And, conversely, it is the investment appraisal process that constitutes the main vehicle for making such plans operational.[4] More will be said about these plans, or strategies, in Section 7.2.

However, the idea that one can conduct an entirely rational search by reviewing the performance of the business, looking at opportunities and deficiencies, and by then triggering investments to address these, is a little optimistic.[5] For a start, we know that few systems that measure performance – that is, accounting systems – are much good at pinpointing difficulties, let alone new markets or the potential for new technologies. A firm will, at any time, only ever have an imperfect understanding of its condition. Second, we know that this kind of search is cumbersome: a small number of people have to gather a large amount of information, often over a lengthy period, and usually at discrete points in time. Business analyses are conducted maybe every two or three years, while the world moves on during the intervals.

What we want is a more adaptable way of identifying worthwhile developments, one more in tune with the idea that a firm is continually changing without ever reaching a certain, planned state. We also want a way of reflecting local circumstances in particular parts of the factory as well as aggregate circumstances across the factory as a whole.

The obvious approach is simply to let people who participate in the company's operations identify opportunities themselves. And the way in which they will do this is by coming upon interesting associations between operating structures and process technology. That is what is happening when the more alert users of simple tools like spreadsheets begin to adapt them in unexpected ways to match particular facets of their jobs. For the most part it is not the knowledge of the technology or of the operations to which it is applied that alone reveals these opportunities. It is the combination of the two. And the combination provides not only the knowledge but the motivation, for there is always a sense of satisfaction to be had in applying a new tool to an old problem.

It is normal to think of the search process in particular, and the planning sequence for new technology in general, as being a centralized affair. But it seems perfectly sensible to say that a part of *any* person's job should be to look for ways of improving that job. He should have the knowledge to be able at least to initiate developments, if not to carry them through in detail. There will continue to be a need for co-ordination, and for the resolution of conflicting demands on limited resources, but this comes after the search.

The fact that a good many developments in the past have come unstuck, not by failing in a technological way, but because of their inability to have much effect on the commercial effectiveness of firms, encourages some people to dismiss technology-led work out of hand. But it is not the mechanism in which the ideas for new projects originate that much affects their outcome. It is the effectiveness of the later stage during which the full range of a development's costs and benefits are assessed, and a decision formed. The main concern within the search process is to maximize the likelihood that valuable opportunities are identified, so that they can be inspected at closer range. It pays to keep the search process a broad and responsive one, and to incorporate all three elements mentioned here (Figure 7.2).

As a matter of interest, it seems that Japanese firms are much more willing than those in the West to spread knowledge about the investment appraisal process around their organizations. Some of the basic principles of financial appraisal are, apparently, taught to almost all quality circles.[6] This is not intended to make everyone in the factory first and foremost an analyst, but to raise the level of financial consciousness. It provides the motivation to look for financially worthwhile opportunities, and it provides the tests that people can apply informally before launching a full proposal.

Finally, it sometimes has to be accepted that the activity of gathering information

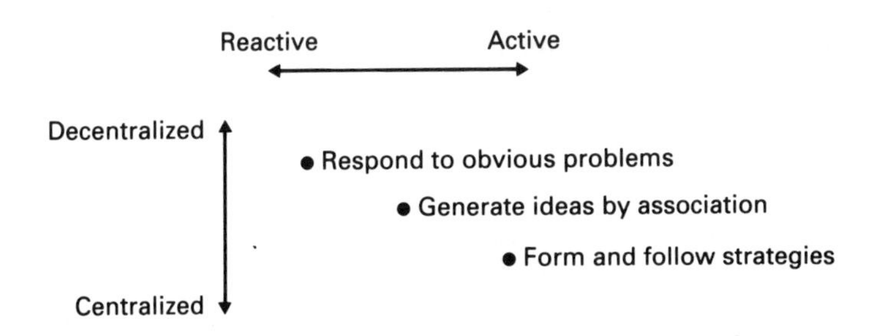

Figure 7.2 Ways of conducting the search process

to help direct a firm's future development isn't always a purely analytical one: pen and paper aren't enough for a comprehensive search process. It is often necessary that a firm commits itself to some sort of action to stimulate the world into revealing the information it needs. It may well be the case that no amount of thinking, research or analysis of past events is going to tell a firm whether its customers would value a shorter product life cycle very highly. In such a case a firm could only find out by experimenting – by introducing the technology and techniques that help it get its new designs to the market more quickly. Similarly, it is unlikely that every material aspect of a technology will become apparent during the planning and appraisal processes. There are some phenomena that can only be experienced to be appreciated. Herbert Simon has said[7] that, very early in the computer era, he advised firms not to acquire computers until they knew exactly how to use them and pay for them. He soon realized that this was bad advice, and that computers initially pay their way by educating large numbers of people about computers.

The decision process

According to Peter Drucker,[8] any decision of an entrepreneurial kind embodies a number of common elements:

- a set of objectives;
- a set of assumptions;
- a set of expectations about what will happen in the future;
- a set of alternative courses of action;
- the choice between one course of action and another;
- the structure of the decision (its position in a sequence of several decisions, and the way in which one restricts the freedom of another);
- a phase in which the impact of the decision is experienced; and
- the set of results that follow.

Any elementary model of the decision process (it could but needn't be Drucker's) is likely to provide a good way of bringing to the surface the terms of a decision about whether to adopt a new development. Writing something down against each of these headings will help decision makers to marshal their data, and help them to hold the important issues at the front of their mind. It will provide an *aide-mémoire* and lend a number of rather intuitive processes a degree of consistency. It might serve to explain the thinking behind it to all those who would be discouraged by any apparent arbitrariness in their managers' behaviour. And it may help to force out muddled reasoning; putting one's thoughts into words is always a salutary activity.

Suppose, for example, that a decision is to be made about whether or not to invest in automating the product handling in an assembly cell – perhaps with some bowl feeders, a robot, and a rudimentary vision system for finding the two-dimensional orientation of incoming piece parts. An appraisal will presumably have been conducted by the time the proposal reaches the decision maker's desk. Probably the proposal has a quantified net present value, and a growth option value that has been described in words but not translated into figures. Our decision maker might now attempt to identify Drucker's eight elements.

(1) The objectives are perhaps to improve the accuracy and repeatability of the assembly process and to explore the applicability of robots in his factory. Underlying these objectives is a broader purpose of course: probably that of maximizing, or at least increasing, the economic value of the firm. But this principle hardly characterizes a specific decision, and on its own it would be superfluous and banal.
(2) A next step is to outline the assumptions on which the decision will be based. Here, the decision maker might be taking as read the fact that the quality of assemblies is more important than their unit cost. He might also be assuming that proven, proprietary systems are available to perform the image processing work satisfactorily.
(3) The following step is to list his expectations of what will happen in the future: perhaps that an existing (or similar) product line will continue to be supplied for the following three years or so, and that this type of assembly process will not in that time be sub-contracted.
(4) He will then list the alternative courses of action open to his firm. These plainly ought to be the same as those considered during the evaluation, as otherwise there is no satisfactory basis on which to compare them. But it may be the case that in finding alternatives that have not been assessed during evaluation he needs to call for a further iteration through the evaluation process. In the assembly cell example, it may be reasonable to consider partial automation as an alternative both to full automation and to doing nothing.
(5) The next element – the decision itself – is simply the selection of a specific course of action.
(6) The structure of the decision is harder to record. In the final analysis, most decisions have some sort of association with most others, however tenuous, and the process of identifying a distinctive sequence or pattern of decisions turns on being able to isolate a particular line of development. In our example, there may be a number of decisions directed towards achieving high and reliable throughput in the manufacture of a specific product line. The decision about the assembly cell may follow one which has established a budget for throughput improvements, one which has selected the assembly process as being the area most in need of attention, and one which has chosen robotics as an important technology for exploration. The assembly cell decision may in turn be followed by decisions about adopting robotics elsewhere, about tackling the design of components for assembly, and about marketing any improvements in the properties of the final products – that is, telling potential customers of the changes (Figure 7.3). Equally, however, the adoption of automation in the assembly cell might fall into a sequence of decisions that are aimed at progressively reducing the size of the labour force: or a sequence of decisions made to get a government grant of some sort. Understanding and writing down the decision structure, as Drucker calls it, ought to be a revealing process. If there is no evident structure

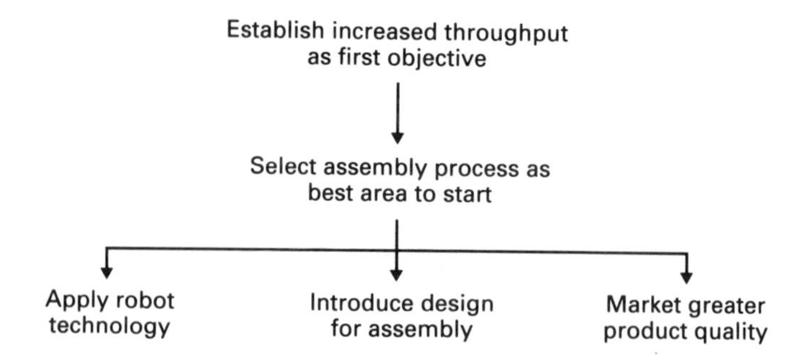

Figure 7.3 A pattern of decisions

whatever, there may be grounds for thinking that there is a certain randomness in the way the firm is being managed.

(7),(8) The final elements of a decision are the impact phase and the results. Both, again, should have been examined in detail during definition and evaluation, but it may help that they are summarized by the decision maker. In the example, the impact phase might be summarized by recording the way in which the automation project calls on resources over its life-cycle: cash flows, engineering hours and so forth. This is really, of course, the province of project planning. The results of the decision ought to be detailed in the evaluation, but it may pay to review the cause and effect networks described earlier. The credibility of the evaluation is as much an issue as its contents at this stage, and that credibility lies above all in displaying a convincing grasp of causes and effects.

Because decisions cannot be isolated from one another, there is a sense in which a decision about a single investment is also one about a bigger pattern of actions. The decision about automating a production process may at the same time be a decision about whether a company should explore automation at all. It is as though there were two decisions that were being made at once – one about a specific investment and one about a general policy, rather than a single decision (on investment) reflecting another made earlier (on policy).

This has the disadvantage of confusing two, distinct choices. It might mean, even if an investment is terminated on the grounds only that specific circumstances make it undesirable that people take this as a signal not to search for new opportunities of that type. If the assembly cell isn't automated, automation might as a result be ignored generally. Equally, of course, a disreputable technologist could swing the case in favour of a project by arguing that the firm is making a policy decision when it decides whether or not to proceed with a project. And the fact that it is a policy decision should override a shortfall in economic value.[9]

A further complication arises because the decision-making process is usually diffuse and sequential. There is no single decision maker, and no single point at which even a group of people makes a specific choice. Momentum tends to build behind a proposal over a finite period of time, involving several people, each choosing to lend their support in a certain sequence. And the activity that precedes the formal sanction of a development encourages expectations and draws out commitments well before the people who officially make decisions get a chance to make them. This is true particularly during the period when information is being collected. King[10] in fact suggests that what is called the decision stage of the investment sequence is less about determining whether a development goes ahead, than it is about formalizing managers' commitment to it. It is not so much about choice as about binding.

The distinction that in practice is drawn between achieving momentum, and officially sanctioning a proposal, means that some care is needed when a model along the lines of Drucker's is applied. The model is plainly concerned with the substance of a choice, rather than its official form. So we might expect it to cover several elements of the appraisal sequence, and not just what we have chosen to call the decision stage.

7.2 Strategic arguments

The discussion in Section 7.1 referred to the use of a strategy for directing the process of consistently searching for opportunities to apply advanced manufacturing technology. Forming and following strategies does much more than direct the investment search, of course, but this is a good point to make some suggestions about their purpose and content.

When strategies are worth forming

The strict definition of what constitutes a strategy, and what distinguishes it from other types of predisposed action, is I suppose rather debatable. But one approach is to say that a strategy is a pattern of actions that are intended to attain certain, stated ends. It might be contrasted with a *policy*, which is more a static guideline for influencing repeated decisions.[11]

There is an obvious contrast between strategic behaviour, which generally follows a sequence laid out at a previous time, and opportunistic behaviour, which is mainly governed by the perception of circumstances as they arise. An opportunist's actions are likely to be directed towards achieving a particular purpose, such as maximizing net present value, and he will therefore test his imminent activity against this purpose. A strategist will, in addition, test his activity against a plan: he will have some preconception about the means of attaining his purpose. A strategist's expectation is that by looking ahead, and by adhering to a consistent theme, his rewards will be higher than if he behaved as though each of his actions took effect independently of one another.

The fact that a strategist is not supposed to pursue attractive side-lines when they might distract him from his strategy means, in theory, that the set of opportunities available to him is restricted. It is always possible that a side-line could prove more fruitful than one stage in the strategic sequence of activities. Perhaps a strategist might have to forgo an opportunity to dramatically reduce costs in one department for the sake of first conducting an audit of the entire factory's operations. An opportunistically-inclined colleague might suggest that following this strategy had lost the firm the accumulated costs the strategy failed to save in that one department.

But in reality, if not in theory, opportunism is a good deal more restrictive than it sometimes appears. Taken too far it easily becomes susceptible to

- short-termism: an excessive bias towards the close future; and
- local optimization: a situation in which sub-systems are made as good as possible regardless of the effects this has elsewhere.

If opportunism is not to become short-termism a firm must have a good deal of knowledge about the future. Exploiting opportunities only on the basis of their individual merits could close off more promising developments later. If, on the other hand, opportunism is not to become local optimization, decision makers must have an extensive and profound knowledge of the present. A cost-cutting technology applied in one department might have some pretty dire effects on others, perhaps.

Moreover, it is not enough that a firm wanting to avoid short-termism and local optimization has a large stock of information. It also has to have a considerable apparatus to process it – to draw the appropriate conclusions. The scale of the calculation needed to assess the impact of a new opportunity in all places and at all times may well prove too demanding.

The information problem is obviously greatest when the opportunity is associated with a technology that is novel and complicated, for there is then much more information to gather. And the calculation problem is greatest when the opportunity is associated with a technology that has wide-reaching and long-lasting effects, for there is then a much more complicated pattern of influences and consequences to trace.

Suppose that at a particular point there is a set X (in Figure 7.4) containing all the possible courses of action that a firm can take. If its managers were opportunistic, they could consider taking any one of these. But if they were to behave in a strategic fashion there might only be a subset of X, known perhaps as Y, of the actions they would contemplate. To be happy about being opportunistic they would want to be sure about two things. First, that they could always know when a particular course of action was in fact accompanied by the risk of short-termism or local optimization. On this assumption, they would reduce their set of alternatives for the opportunistic case from X to a subset Z. And, second, they would want to know that opportunism remains the better approach even after they had made sure of avoiding all options accompanied by short-termism or local optimization. In other words, that Z contains Y, or at least that the contents of Z are more promising than the contents of Y.

If it is possible that one or other of these conditions isn't met, then it is worthwhile examining how a more strategic approach to selecting technological opportunities might be adopted.

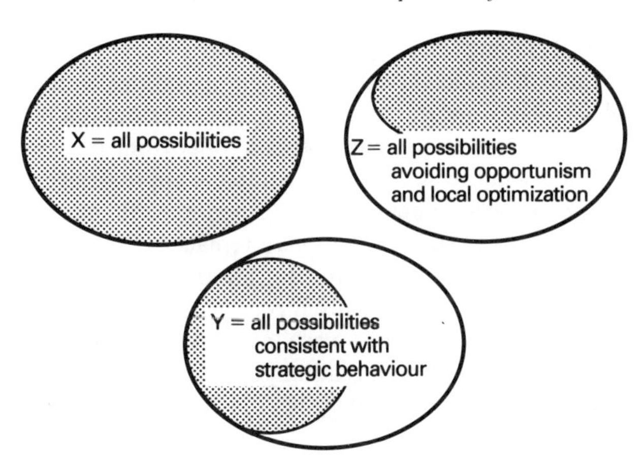

1) Can the developments that avoid short-termism and local optimization (Z) be identified?
2) Is this set a better one than the developments consistent with a strategy (Y)?

Figure 7.4 Strategic *versus* opportunistic developments

Strategies and financial justification

Because, in adopting a strategy, decisions about a bundle of actions are made at once, it has been suggested that the justification and funding of proposals should be conducted at the level of the strategy rather than at that of the individual project.[12] In other words, you evaluate the net worth of the strategy, and if it is acceptable then you can proceed with all the constituent developments without another word being said.

If a strategy is so firm that the subsequent rejection of the single actions that constitute it will simply not be contemplated then this approach has some appeal. There is, after all, just one decision and there need therefore be just one evaluation. If, however, the strategy is only a guide to future action – if it is a weaker thing that is directed mainly towards the search process described earlier – then this approach looks less appealing. The particular difficulty is getting and processing all the information needed to assess a complete strategy. The information problem is difficult enough as it is with single, well-defined developments. So justification by strategy is an interesting thought, but its use ought to be considered with some care.

In practice, bundling up a series of investments in one package and calling the result a strategy, is sometimes a convenient way of getting the weaker of the investments sanctioned without having them exposed to a rigorous analysis. Assuming that a decision maker wants to avoid this, he needs to test the connections between the constituent investments. If there is any sense in which they are independent, it makes sense to call for a separate appraisal for each.

Finally, it is worth saying that some commentators[13] contrast companies which impose a high degree of financial control on their operations with those that adopt a strategic emphasis. The important distinction, however, lies not between financial control and strategic operation, but between opportunistic and strategic approaches. It is quite wrong to suppose that strategic approaches necessarily stress operational factors or technologies at the expense of financial objectives and financial management. A particular firm *may* replace the financial evaluation of individual projects with a strategic plan. But another firm may equally demand that not only does a project individually meet a financial criterion, but also that it is consistent with a finance strategy – perhaps one geared to reducing fixed costs. In other words, a strategy can be anything you want it to be, and its focus can lie in any particular place you want it to lie – including the financial domain.

The benefits of pursuing strategies

The process of forming a strategy is yet another call on scarce resources. And since it rules out a degree of opportunism there is a finite probability that worthwhile possibilities are missed. In other words, strategies have opportunity costs, and it is as well to be clear about the benefits they offer. The purpose of this sub-section, then, is to suggest what these benefits might be. The list will not be comprehensive, but it might prompt some thoughts. Suppose, for the time being, that we are considering only strategies in their weak form: that is, as guides to the search process rather than as fixed intentions.

First, strategies can economize on information. By following a strategy, the people who look for new opportunities to invest in advanced technologies, gather data on them, and produce forecasts, can discard most possibilities on the grounds that they don't conform to the strategy. Moreover, they can probably be more cavalier with the proposals they do feel they have to evaluate. The fact that they follow a strategy while behaving in such a way suggests that it is more likely that they will make sensible choices than if they followed a process of purely random selection.

Strategies are, to some extent, models for an organization. Instead of exhaustively assessing itself and the outside world, the adoption of a strategy suggests that the firm is imposing on the world a conception of how it wants to be. It is following a self-fulfilling plan. While those who form strategies probably take account of existing conditions, they are essentially expressing a pattern to which they expect their company to conform.

The sense of conscious positioning that comes with the application of a strategy can also help secure options. Given that single developments never proceed exactly as anticipated this is an important way of mitigating uncertainty.[14] Of course it also implies that strategies formed on the basis of a firm view of how events will unfold in the future are likely to be flawed: one must consciously build into the strategy the determination to acquire options that will help the firm deal with specific areas of uncertainty. An example might be to stipulate that a new computer-aided engineering system should be able to cope with certain types of change in the nature of the firm's products – from electro-mechanical to electronic, or microelectronic, perhaps.

A further benefit that ought to accompany the use of a strategy is better motivation. A strategy communicates something most people don't hear about during the daily grind: their place in the whole, and its future development. Such a communication can allay fears about the future, and it can have a more positive value in giving people a sense of direction. It connects a person's job of today with the jobs of others, and it connects it with his job of tomorrow. It might be especially important to him when he perceives a threat to the job of tomorrow, maybe because automation is being introduced. It might also be important when he is called upon to change his job for the sake of improvements that can only be detected elsewhere. How many draftsmen, for instance, would feel enthusiastic about CAD without being confident that those who introduced it could say how it would affect the draftsman's job, how it affected those who made use of his work, and how it fitted into the company's expressed intention to improve its return on capital?

So it is reasonable to suppose that forming strategies, and telling people about them, can improve both focus and drive: focus by stating goals and being clear about how they are to be reached: drive by indicating how individuals and groups play their part in all of this. It is difficult to know in quantitative terms how much this sort of thing can improve motivation, and how much motivation can improve aggregate performance. But you might be able to get a feel for the importance of the latter by looking at some of the literature on what has been called X-efficiency.[15]

Another benefit that ought to flow from the adoption of a strategy is the ability to co-ordinate developments in distinct departments. This is especially necessary when those developments incorporate information technologies: these tend to have widespread effects, and they are notorious for failing to respect departmental boundaries. Information systems have a much greater reach than mechanical systems, and, when they install them, companies lose some of the human buffering between one system and another. This means that a degree of compatibility between the output of one system and the input specification of another is very important. It is perfectly possible to transform the outward form of information as it is transcribed between two systems, but it may also be necessary to *add* information during the transcription. If the meaning of one system's output is narrower than the meaning of what another system needs to know, there are likely to be problems.

Of course duplication is as likely as omission. When a firm's managers lack a clear understanding of how several sub-systems are to work together, and develop in a compatible way, it is not unknown for more than one sub-system to be doing the same job. This is usually wasteful and occasionally amusing, for it means that different parts of the organization work on the basis of different assumptions about similar phenomena and events. Again, a way of avoiding such problems is to follow a strategy of some sort.

Lastly, because a strategy is *the* express understanding of how several developments are interconnected, it is an important opportunity to forge links between different disciplines. It is a way of getting one specialist to take an involvement in another specialist's problems. For instance, a firm might choose to pursue a strategy intended to halve the total lead-times on its contracts. It will doubtless attempt to understand the effects in commercial terms (how appealing shorter lead-times are to customers) and in financial terms (the increase in revenues that follows, offset by any increase in costs). It will probably identify a series of actions designed to improve operations. This might include the introduction of concurrent engineering, the adoption of design-for-manufacture policies, and so on. It will probably identify the technologies intended to help bring about these operational changes. There is little question that this process of translating influences from one language to another needs the participation of people from different disciplines: of finance managers, commercial people, production engineers and computer analysts, for instance.

There are therefore at least three good reasons for forming and following strategies:

- to economize on information;
- to improve motivation; and
- to achieve co-ordination.

This gives us a basis for choosing among different strategies and for testing the effectiveness of any one of them.

An example

The example that follows ignores many of the complexities of the real world – the intuitiveness with which people approach problems, the way groups of people reach points of consensus, and the short term fluctuations in a firm's fortunes that cause it alternately to drop and resume its pursuit of a strategy. It is simply intended to illustrate some of the points made in the previous sub-section.

Suppose that we are attempting to form a strategy for the introduction of advanced technologies in order to improve our firm's engineering and manufacturing processes. As it is not a marketing strategy, it is reasonable to assume that products and markets are already determined. Similarly we shall choose to ignore any effects our technologies might have on market structures – barriers to market entry and so forth.

For the sake of argument, our explicit goal will be to maximize our firm's economic value. We shall, in other words, seek and exploit opportunities to increase the net present value and growth option value of future cash flows. Although the individuals within the firm will have their own personal objectives, it is reasonable to expect them to accept the legitimacy of maximizing economic value (even if they sometimes don't like the implications). In this particular case we shall concentrate on benefits in the firm's product markets – perhaps because we have heard that advanced technologies have a greater influence on revenues than on costs. So we first need to consider how to interpret such a purpose, initially expressed as a financial effect (that of increasing revenues), in commercial terms. This means understanding the things that influence our customers' belief that our products are cost-effective and preferable to those of competitors. Although we have assumed that product and markets are predetermined, we still need to know something about them.

There are two views we might take of how potential customers form opinions about our products and make choices as a result. The first is that they use

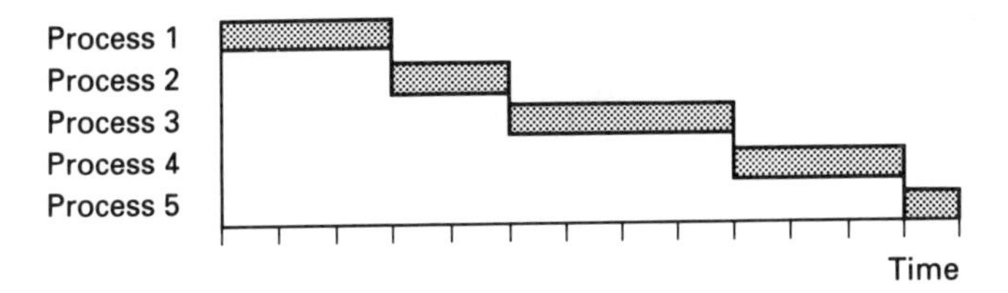

Figure 7.5 A product's progress

compensatory models (much along the lines of the scoring models described in Chapter 4). Perhaps they reason that a property such as functional ability is twice as important as a short delivery period, but only one-and-a-half times as important as quality (that is, fitness for purpose). According to this model, the potential customers score their options on these scales, compensating the effects on one with those on another.

The second view is that they use non-compensatory models, the most obvious of which is a simple ordering. In other words, customers have a most important property – such as functional ability – by which they first compare their options. Only if the options are roughly equal in terms of this property, within the applicable bounds of uncertainty, would the next most important attribute (fitness for purpose) be examined. The process continues until a distinguishing attribute has been found.

Suppose we adopt the view that customers choose products according to a non-compensatory model. In decreasing order of importance they consider sequentially a product's functions, then its quality, then its delivery time and so on. Suppose also that we assume that our product's functions are fixed as far as a process technology is concerned, and that quality has reached the stage throughout the market where it is difficult to much improve our position by tackling it. Our commercial understanding would then indicate that we ought to focus our strategy on reducing delivery times. If ours is an engineering firm, producing capital equipment perhaps, delivery times can be equated with lead-times over both engineering and manufacturing activities.

Before looking at a technology strategy we plainly need to interpret the commercial intention as an operational one. We need to understand how our operations need to be developed to achieve shorter lead-times. We could make a start on this by drawing a typical schedule for the design and manufacture of a product, of which Figure 7.5 is a conveniently trivial representation.

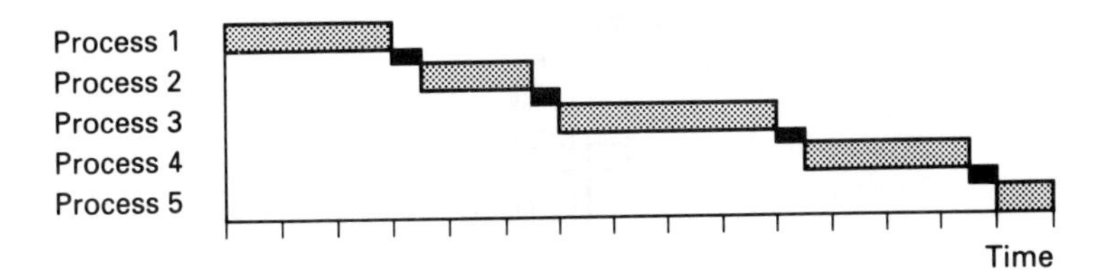

Figure 7.6 Handover delays

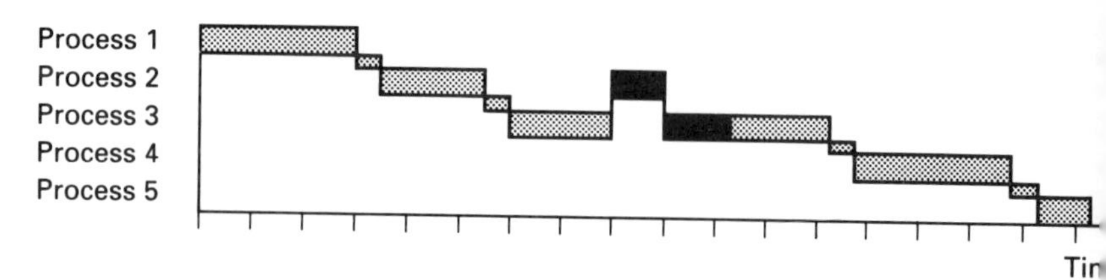

Figure 7.7 Rework delays

Now we know in reality that the hand-over, as it were, from one department to another tends to be rather fraught, so a more realistic picture of the schedule might turn out to be that of Figure 7.6 (in which the extra times introduced by handover delays are drawn in solid shade). Even this tends to be optimistic if we regularly suffer rework for one reason or another – perhaps because of upstream errors in areas like design and drafting, poor raw materials and so forth. This makes the schedule still less satisfactory (Figure 7.7)

It is evident from this schedule how we might develop the firm's operations in order to reduce lead-times. The most obvious things to tackle are rework and handover problems, and to try to reduce the time taken for each process individually – provided that we don't at the same time make rework and handover problems any worse. But one of the main things we can also attempt to achieve is concurrency. On the shop floor this can be done by reducing batch sizes: since process n may start its work on a batch when process $n - 1$ has completed its work on the same batch, the smaller the batch the earlier process n can be started, all other things being equal (Figure 7.8).

The idea of homogeneous batches is inappropriate to most engineering activities – product specification, product design, process planning, fixture design, part programming and so forth. But we must clearly attempt to make the output of these activities available to processes that succeed them as quickly as possible. Just as soon as the full complement of inputs to a process have been prepared, that process should begin. People should not have to wait until a previous process is complete if what they need from it becomes available *before* the process is complete.

By this stage, we have an understanding of the operating issues that will need to be addressed: concurrently acting processes, small batch sizes, slick handovers, and reduced rework. We might care to attach some numbers to the extent to which

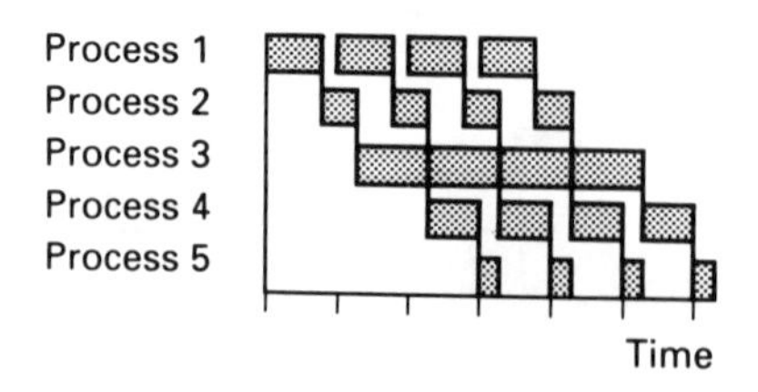

Figure 7.8 Small batch concurrency

we want to reduce lead-times. It would certainly make sense to estimate the extent to which each operating improvement contributes towards reducing the product's total lead-time – this will help order our priorities. And we can now consider the technologies that appear promising. The ability to work with small batch sizes suggests a need for shop-floor systems geared up to just that: systems that will minimize transportation between processes, that will help perform the complicated scheduling and dispatching which will in turn be needed to achieve synchronization and so forth. In some areas we might concentrate on techniques, such as cellular groupings based on product lines rather than production processes. In others we would need to consider buying technologies such, perhaps, as OPT scheduling programs.

To help achieve a measure of concurrent engineering, we need to trace the information that enters and emerges from each identifiable process. From that we ought to be able to work out the necessary precedence between these processes. In particular, we need to understand at what point we can initiate any process that is dependent on others reaching a certain stage. To help our information flows we could consider buying an engineering data repository so that as soon as a piece of information is ready to be passed from one process to another it is easily accessible. It would be natural to look for a repository that allows us to define quite clearly when a piece of information is incomplete, and when it has become complete and ready for use.

To tackle handover problems we need the same sort of thing: clear, unequivocal control over the conditions under which information passes from one process to the next, and easy access to information from wherever an engineer happens to be sat. Again, we can look to our engineering data repository as a suitable technological aid.

We might also want to involve the people responsible for a later process in the execution of the earlier processes on which it relies. Rather than the handover being a lengthy negotiation and transfer of information in which that information might have to be revised, we then have a formal handover for which all the technical difficulties associated with one discipline taking over the products of another have already been resolved. In other words, we need an approach to design-for-manufacture.

This will also help us to tackle rework. One process should no longer, in principle, receive the products of a previous process which are in some way deficient because the handover was muddled – maybe because the requirements of the later process were misunderstood, or the products released too early. In addition, it is probably worth looking at the quality of each process individually to see how far any recurring types of error can be eliminated. Technologies that are, above all, aimed at getting things right first time are worth considering here. For example, a knowledge-based system that helps industrial engineers choose the right tooling might be worth exploring. Irrespective of whether the system knows more than the engineer, it can help make sure that all factors relevant to the choice of tool are properly taken into account. It can also help to make sure that difficulties experienced in the past by any engineer are made known whenever they are relevant.

At this point, we have an idea of the techniques and technologies that we should consider, and those that are likely to prove the most worthwhile – and therefore those that deserve higher priorities. In addition to a set of priorities based on the scale of the expected impact, we ought also to consider the pattern of precedence between the various developments. We might be especially concerned that organizational changes, and questions of technique, precede the installation of technology. The result will be a pattern of developments that indicates where we should start, and what should follow what.

This is not quite enough for us to be happy that we have yet produced a promising strategy, because there is still some indeterminacy in the pattern of the individual projects. If, for instance, we cannot be certain which of actions *X* and *Y* would prove most beneficial, and if neither depends on the other in any way, we do not have any criteria, so far, to help us say which should come first, or even whether they shouldn't be undertaken at the same time. It would therefore be helpful to estimate, if only in outline, the resources that will have to be committed to each action. Matched against existing resources, this will give us some idea of when developments will be practicable.

And that, except for the detail, is sufficient to make this a strategy, although it provides little more than a crude guideline for the search process. The proposals for specific developments that emerge have still to be thoroughly appraised to determine whether they promise to have a positive net worth.

There are a number of slipshod elements in the process just described. The developments which have the greatest impact on revenues need not be those that yield the greatest *net* benefit after their opportunity costs have been considered. It is not clear whether the strategy has the weak form of a guide to the search process, or the strong form of a plan of campaign. Moreover, I have simplified certain elements to the point of triviality. Assuming that we can focus on a single, highly distinct product line in complete ignorance of others would be a nonsense in most firms.

There is no particular reason why a firm should have an advanced manufacturing technology strategy at all. It might prefer, for instance, to have an information strategy – one in which it decided how it was going to manage information throughout its operations. And strategies do not necessarily originate in the way I have suggested here. The more conventional approach is to develop strategies by reconciling a firm's characteristics (its strength and weaknesses) with expected developments in its surroundings (the threats and opportunities).[16] The important thing for a particular firm is evidently to adopt an approach that will, in its own special circumstances, stand a good chance of revealing the strategies of greatest value. The tests of a strategy's value – information economies, higher motivation and adequate co-ordination – have a wider applicability than specific methods of forming one in the first place.

All the same, the example illustrates one way of approaching the formation of strategies. This approach is based on the idea that objectives tend to be framed in financial terms, that commercial issues play a large part in determining how financial objectives can be met, that operational issues have to be considered in any attempt to improve commercial performance, and that process technologies are a way of tackling operational issues. The chain of consequences is not necessarily this extended, especially where technology can be applied directly to a firm's

Figure 7.9 The usual direction of impact

commercial processes – selling and buying transactions for instance. But it is the pattern that seems most probable when we are considering manufacturing technologies (Figure 7.9).

Some other considerations

There are a few additional issues that need to be thought through when adopting a strategy. It is rarely obvious, for example, whether it would suit a firm better to lead or to follow in applying new technology, and it is useful to know which it is in fact doing. It affects the level of risk to the extent that information about the appropriateness and simplicity of a technology can pass (whether deliberately or not) from a leader in its application to the followers. It also affects the structures of markets in which customers attribute a measure of goodwill to suppliers that visibly attempt to innovate. In other words, the technological abilities of competitors can affect the benefits a firm gains from introducing new systems.

Using a strategy is a form of control every bit as much as the act of making detailed operating decisions. As such, it is likely to be ineffectual without some grasp of how its success is to be measured. It is hard to take such measurements when the company can never (as it shouldn't) reach a position of stasis, for there is always some new development coming on stream, and some new external factor beginning to take effect. In fact we might wonder how measurements can be taken at all that would distinguish long-term effects from random, short-term fluctuations, and controllable influences from uncontrollable pressures. Even if this were possible, people often have incentives to misrepresent or hide the findings – typically to attribute failures to the environment and successes to themselves. It is difficult to find any principle more profound than the idea that people should be faced with the consequences of the plans they compile and the decisions they take.

It makes sense to tie in technology strategies with all the other types of strategy that seem to abound – manufacturing, marketing, product strategies and so on. There is a suggestion,[17] for instance, that in the earlier parts of a product's life-cycle the order-winning criterion is usually a low price, which suggests a need above all for cheap manufacturing processes. Later in the cycle, by this account, the typical order-winning criteria become delivery speed and reliability, placing the stress more on flexible processes. In contrast, others[18] think that new process technologies set a trend against price-based competition generally. Instead they emphasize market segmentation, and competition on the basis of customizing products to the desires of individual customers. So there is clearly no consensus on how technologies line up with marketing strategies. But at least it's more defensible to have a consistent set of strategies than to have a set of contradictions and *non-sequiturs*.

Finally, strategy formation is (luckily for those who conduct it) a continual activity. New factors and new technologies enter its scope and old ones leave from time to time. Against the background of a more constant purpose, perhaps that of maximizing present value, there is a changing interpretation of how it is best achieved. Although a certain sustained focus is needed to achieve anything that is planned for the future, it is generally accepted that firms should not attempt to look for final solutions to their problems. The outside world moves on in ignorance of one firm's intentions for it.

7.3 Real decision-making behaviour

Maximization as description

The principles of making decisions about industrial investments are usually normative in character: they prescribe a supposedly ideal way of exploiting opportunities but they do not attempt to explain how decisions are taken in practice.[19] This ideal is commonly taken to be that firms always act to maximize economic worth, a quantity that we would here equate to net present value plus growth option value. The decision rule mentioned in Section 5.1 was intended to capture this principle. Managers are always meant to adopt the course of action that is associated with the highest anticipated worth. Sometimes constraints on resources such as money and expertise are explicitly recognized. But even then, one is to regard economic worth as the single quantity that encapsulates everything that is of value to a firm – and to maximize it. The point, for instance, of using the kind of strategy discussed in the previous section was to locate the *best* opportunities to apply new technology, not just to find opportunities that were merely adequate. The point of measuring them against the economic value yardstick was to determine which of them represented an *optimal* course of action. Both the search and evaluation processes were geared towards maximization.

There are several reasons why this maximizing rule is almost never observed in practice, and some of these also bring into question the idea that the rule should be regarded as an ideal worth aiming for (even if it is never achieved).

To begin with, the appraisal process has a limited number of resources at its command: it can only collect and interpret a fraction of the information that is relevant to its operation. This became apparent in Chapter 5 as soon as we started thinking about how the effects of a new technology could be translated into expected cash flows. If we had wanted to build an analytical model of the entire pattern of relevant causes and effects we would have had both to gather a great deal of raw data, and to perform a large amount of manipulation on it before it yielded the cash flows and risk factors we needed for a present value calculation. It was not even clear that the raw data was there to be gathered: most firms have a relatively imprecise understanding of their own operations, and they often lack experience in applying new technology. We can't, however, claim to be able to maximize our firm's worth if we don't have the knowledge that reveals the best among all the available courses of action. You could argue that it is still possible to maximize within the constraint of the information that we do have, but of course it is always possible to gather more information (at a cost). As value maximizers we would then have to calculate the optimal amount of additional information we should collect, a calculation which needs still more information.

Another reason why maximization is a poor description of real industrial behaviour is that there is invariably conflict within organizations. Different people are simply intent on pursuing different objectives. Since it is usually plain to see that much of this conflict remains unresolved, it is unrealistic to suppose that a firm will ever pursue a single goal with any constancy, or even pursue a coherent compromise. What is much more likely to happen is that the firm will attend to goals sequentially,[20] pursuing one for a while and then another without ever attempting to knock them into one higher goal, such as the value maximization principle. If goals are pursued sequentially, at different times, they need not be consistent with one another: marketing and production goals, for instance, have traditionally been in conflict. This hasn't stopped firms alternatively veering from one to the other.

It also seems that in practice global goals such as maximizing value are just too abstract. They are too far removed from day-to-day operating decisions to be tractable,[21] and we can expect managers and workers to pursue more parochial, but more tangible goals – such as reducing work-in-process, increasing market share, cutting engineering costs and so forth.

The result is that instead of attending to the value of a single quantity (such as NPV), real decisions are made as though managers consider the impact of a new development on a variety of quantities. Variables such as turnover, accounting profits, the size of inventories and so forth might represent the type of quantity which managers use as proxies for more fundamental notions of value. Moreover, instead of maximizing these quantities, they appear to be content simply with raising them above particular thresholds. Managers have a set of aspiration levels, as it were, and they are happy to adopt a particular development if it promises to meet these levels.

The nature of the search process, as well as that of the evaluation, now changes. The search for potential developments commonly begins only when a clear problem has arisen. It only continues until the first course of action that appears to meet the aspiration levels has been found.[22] The order in which the search is carried out therefore strongly influences the developments that are in fact adopted. And tests that are applied to such possibilities are tests of feasibility rather than of optimality – they are designed to discriminate between the sets of acceptable and unacceptable opportunities, not to identify a single, outstanding opportunity. This is plainly very different from the idea of a firm continually seeking new opportunities and optimizing its choice of development to squeeze the greatest possible value out of its activities.

Maximization as prescription

These reasons are not enough on their own to make maximization suspect as an *ideal* form of behaviour. There are at least two other factors, however, that might be enough. The first is that a world of cash flows and discount rates is inevitably a highly simplified one. It is an abstraction of all the conditions and connections in the real world, and it is an expression of a particular scheme of values. It is not at all obvious that optimizing one's behaviour according to such a simplified model is any better than behaving in a merely satisfactory manner against the background of a more realistic model.[23] We have already seen that with new technology there is a danger of ignoring effects that do not arise in monetary transactions. These effects are often material to the appraisal, and to find an optimal solution to a model which ignores them might be to find a wholly unsatisfactory solution.

Second, when the environment is relatively unstable – especially when it is unstable in an unpredictable way – a type of myopic, adaptive behaviour might be more robust than one based on long-term planning.[24] In other words, in conditions of great uncertainty, one attempts to survive by quickly matching one's actions to outside influences. Predicting these influences long in advance and forming complex, slowly-acting strategies can render firms vulnerable to unexpected perturbations in their environment. An attempt to maximize economic value may be much less appropriate than maintaining short-run liquidity, perhaps, when this is the case.

This line of reasoning can, however, be taken too far. Adaptive behaviour, in which managers pay attention to aspiration levels rather than a single, global goal, can blind them to fundamental changes in the world at large – changes which call for fundamental adjustments to the criteria by which a firm measures its performance. As we saw in Chapter 2, much of the justification for new technology lies in the changing patterns of demand in product markets. This suggests that aspiration levels based on an earlier phase (such as one in which the main intention is to achieve a low cost base) will become misleading over time. There is also a danger of concentrating exclusively on external changes that register in short periods of time, and completely missing substantial but more slowly-acting developments.

Also, short-run decision making is based on working within existing constraints. Big issues that cannot be tackled quickly are treated as uncontrollable limitations on the actions that a firm can take. For example, one might regard the cost of a certain manufacturing process as being such a fixed constraint. But reports of the way in which some Japanese firms are managed[25] suggest that these firms are ready to tackle such constraints, and the tests they apply to new developments are not simply tests of feasibility. This behaviour has the character of maximization, and it may well be that competitors will have to emulate this in order to prosper in the same markets.

Strategic management is also less a case of making assumptions about external factors than it is of forcing the firm to take a particular form. Many of the plans it uses are self-fulfilling, in the sense that they can be made to happen by taking actions that are within the control of the firm's employees. These actions are therefore relatively robust to changes in the environment. One could pursue a strategy designed to achieve economies of scope, or to manage elements of the organisation by a type of client-contractor relationship, and the results of such a strategy would probably be insensitive to economic fluctuations outside the firm. It is untrue to say that a pattern of long-term intentions is necessarily sensitive to the assumptions on which it is based. It is, as a result, questionable to claim that a firm's survival is necessarily better served by short-run, adaptive behaviour than by long-run planning. In practice, a combination of both is perhaps the most sensible approach.

Technology and aspiration levels

If we took the view that at least one element of the decision-making process was based on meeting aspiration levels, rather than on maximization, how would we assess a new technology?

We would first want to identify a number of fitting aspiration levels, then gauge what influences them, and finally determine how a new development changes these influences. To a large extent, the set of relevant aspirations will be a function of local circumstances. But an obvious ambition of almost all firms will be to maintain their economic worth intact. That is, decisions shouldn't be taken that lead to a depleted combination of earnings and earning power. Now in the form of net present value and growth option value we have a satisfactory measure of this combination, and this suggests that a sensible aspiration would be that any new development should promise a sum of present value and option value that is greater than zero. If a proposal that promised a negative value were ever adopted, there is plainly an expectation that the worth of the firm would be reduced.

An aspiration expressed in such a way is still relatively abstract. A certain amount of calculation is needed to check that it is being achieved, and its application might therefore be restricted to substantial, long-lived actions – that is, to capital investments. But the advantages of casting aspiration levels in terms of economic value rather than in terms of accounting profits, turnover, payback periods and so forth are considerable (as seen in Chapters 4 and 5). This makes a certain amount of abstraction acceptable. By focusing our attention on present value and option value we can in particular avoid

- sacrificing long-term growth for short-term growth,
- sacrificing earning power for earnings,
- sacrificing margins for turnover.

By using it as one among several aspiration levels, the *maintenance* of economic worth imposes much less strict rules in the search process than the *maximization* of economic worth. It means that one doesn't attempt to carry out the search exactly up to the point at which the extra expected value of the revealed opportunities equals the extra cost of conducting the search. Such a calculation would of course be impossible to perform in practice. It also allows the evaluation process to pull into its scope additional aspirations that cannot be connected directly with economic value. Minimum standards of human safety, for instance, can be expressed as a second aspiration level: the analyst needn't try to convert increments of safety to positive cash flows.

Aspiration levels commonly change over time. They tend to rise while conditions are benign and fall when they are unfavourable. We can therefore expect that our requirement that economic value exceeds zero will be adjusted from time to time, and that in favourable circumstances we will test for an economic worth that exceeds a positive (rather than zero) threshold. A survival test might become a growth test, and our aspiration level for a new development's economic worth might be such that we demand that it exceeds, say, 10% of the initial investment.

Whatever the particular standard a firm effectively applies at any one time, it can scale that standard using the yardsticks described in earlier chapters. Whether a decision maker acts as though he were maximizing, or whether he simply tries to satisfy a number of aspiration levels, these yardsticks provide a convincing way of expressing an objective system of values. Because the appraisal schemes we have considered do not embody particular decision rules (although they may suggest them), we can continue to use them even when we are uncertain about how decision makers will reach their conclusions. As has already been said, our hope for an appraisal is that it will inform, not automate, the decision-making process.

7.4 Summary

The sequence in which new systems come to be adopted commonly embodies a number of stages:

- triggering – when the impetus to take up an opportunity becomes strong enough to initiate some kind of action;
- screening – an informal filter applied to remove opportunities that are not worth a formal appraisal;
- definition – a stage in which the scope and nature of a development are clarified;
- evaluation – a formal assessment of a development's likely value; and
- decision making – the final choice or sanction involving the commitment of a firm's managers to the development in question.

Screening is an especially important stage to think about because it often operates on the basis of parochial concerns, using experience and rules of thumb which may discriminate against advanced systems.

New opportunities to apply advanced systems arise from a search process. This can work in a number of ways:

- it might be reduced to finding solutions to pressing problems;
- it might be the result of an explicit strategy which generates a number of obvious proposals;
- or it might generate ideas by association – by linking operational needs with technological opportunities.

Search by association is likely to be fruitful with new technologies since they are highly programmable and therefore adaptable to their applications. Strategies, however, are especially important for new technologies. These technologies, because they dispense with the human buffering formerly found in the organization's information chains, make greater demands on its ability to co-ordinate the operation of its sub-systems.

Perhaps the easiest way to characterize strategic behaviour is to contrast it with opportunism, the dangers of which are primarily

- the possibility of short-termism, and
- the possibility of local optimization.

The benefits of adopting a strategy are nonetheless more positive than simply avoiding the risks of opportunism. They may be used to

- economize on information,
- improve motivation, and
- achieve co-ordination.

Strategies for advanced technology will normally acknowledge the extended chain of causes and effects mentioned in earlier chapters. One might therefore go about framing a strategy by

- expressing a financial goal,
- translating this into commercial goals in the firm's product markets,
- translating these in turn into operational goals that specify the conduct of the firm's internal processes, and finally,
- working out how these operational goals may be reached by introducing specific technologies.

These ideas are geared towards finding the best opportunities open to a firm, but in practice the notion that the decision-making activity can be optimized is flawed:

- decision makers don't have enough information about the future to know when an action is optimal;
- internal conflict means that firms don't have a single, consistent set of goals;
- the instability of the environment demands fast adaptation rather than planning; and
- the models on the basis of which it is realistic to optimize are simplistic.

It is often suggested that real decision markers behave as though they were attempting to meet thresholds on a variety of parameters, not maximize the effects on just one. It is also believed that the search for new opportunities is initiated only by pressing problems, and that it is sequential in nature – that decision makers consider only the opportunities they come across before finding one that meets their aspiration levels.

There are dangers in having aspirations and aspiration levels in the advanced firm which were really only appropriate to firms in earlier times. The yardsticks considered earlier, however, are sufficiently general that they avoid the problems of re-using inapplicable rules of thumb and outdated experiences. They provide a way, in particular, of expressing an aspiration to keep a firm's economic value (its earnings and earning power) intact.

Notes and references

1 This and several of the succeeding stages mentioned here are those suggested by King, P. Is the emphasis of capital budgeting theory misplaced? *Journal of Business Finance & Accounting*, **2**(1), 69–82 (1975)

2 Dobbins, R. and Pike, R. The capital investment process. In Ivison, S. *et al.* (eds.) *British Readings in Financial Management*, Harper & Row (1986)

3 Williamson, I. P. Factory 2000: justification in a strategic context. *Proc. International Conference on Factory 2000*, Cambridge, 31st August (1988)

4 Pike, R. H. A review of recent trends in formal capital budgeting processes. *Accounting & Business Research*, **13**(51), 201–8 (1983)

5 Moss writes about a search process influenced by focusing effects (considerations of existing resources and objectives) and inducement effects (the forces of competition forcing innovation): Moss, S. J. *An Economic Theory of Business Strategy*, Martin Robertson, Oxford, pp. 53 and 59 (1981)

6 Fujita, S. Engineering economy in Japan. *Engineering Economist*, **27**(3), 238–40 (1982)

7 Simon, H. A. Designing organizations for an information-rich world. In Simon, H. A. (ed.) *Models of Bounded Rationality. Volume 2: Behavioural Economics & Business Organization*, The MIT Press, Cambridge (Mass.), pp. 171–85 (1982)

8 Drucker, P. F. *Technology Management and Society.*, Heinemann, London, p. 166 (1970)

9 The connection between policy decisions and investment decisions is mentioned by Bonsack, R. A. Justifying automation in the office and the factory. *Journal of Accounting and EDP*, **3**(3), 63–5 (1987)

10 King, P *op. cit.*

11 This distinction is suggested by Wilson, B. *Systems: Concepts, Methodologies and Applications.* Chichester, John Wiley, p. 161 (1984)

12 Michaels, L. T. A control framework for factory automation. *Management Accounting* (US), **69**(11), 37–42 (1988)

13 For example Jones, W. G. T. Company strategy – a role for the electrical engineer? *Engineering Management Journal*, February, 12–8 (1991)

14 Crum, R. L. and Derkinderen, F. G. J. Multicriteria approaches to decision modelling. In Crum, R. L. and Derkinderen, F. G. J. (eds.) *Capital Budgeting Under Conditions of Uncertainty*, Boston, Martinus Nijhoff, pp. 155–73 (1981)

15 For example Leibenstein, H. Allocative efficiency vs. X-efficiency, *American Economic Review*, **56**, 392–415 (1966)

16 Argenti, J. Corporate Planning. *Accountants Digest*, **170**, Spring (1985)

17 Hill, T. *Manufacturing Strategy*, Basingstoke, Macmillan Education, p. 212 (1985)

18 Goldhar, J. D. and Jelinek, M. Plan for economies of scope. *Harvard Business Review* November-December 141–8 (1983)

19 A discussion of this distinction and its implications is commonly a starting point for papers on behavioural principles of decision making. See, for instance, Einhorn, H. J. and Hogarth, R. M. Behavioral decision theory: processes of judgement and choice. *Journal of Accounting Research*, **19**(1), 1–31 (1981)

20 Carter, E. E. The behavioral theory of the firm and top-level corporate decisions. *Administrative Science Quarterly*, December, 413–28 (1971)

21 Simon, H. A. Rational decision making in business organizations. *American Economic Review* **69**(4), 493–513 (1979)

22 Cyert, R. M. and March, J. G. *A Behavioral Theory of the Firm*, Prentice-Hall, Englewood Cliffs (NJ) (1963)

23 Simon, H. A. *op. cit.*

24 Cyert, R. M. and March, J. G. *op. cit.*

25 Hiromoto, T. Another hidden edge – Japanese management accounting. *Harvard Business Review*, July-August, 22–6 (1988)

8 Understanding the risks of technology

The gambling known as business looks with austere disfavour upon the business known as gambling.

Ambrose Bierce
(in H.L. Mencken *A New Dictionary of Quotations*)

8.1 Introductory ideas

Risk and uncertainty

To say that a development is risky is to say that there is a conceivable chance that its outcome will turn out to be somehow different from the one that seemed most plausible or most likely. The predicted effect of installing a new robot might be a more repeatable, more accurate and cheaper assembly process, less rework and fewer warranty claims. But there may be a possibility that the robot fails to meet the tolerances needed, that its sensors cease to work in the dirt of a real factory, or that its programming proves to be too complex for untrained technicians.

The people weighing up the case for buying the robot are faced with such risks because they lack a perfect knowledge of the consequences of their actions. If they knew unequivocally whether the robot would be successful, they would know whether or not to buy it: in either case, they would, as a result, face no risk.

Although the term *risk* has overtones of loss, it is worth pointing out that a simple lack of information is not a purely negative influence: unexpected outcomes might be unexpectedly favourable ones. Considered risk-taking is, of course, the essence of commercial life. In fact the greater the general degree of uncertainty about the shape of the future, the greater the chance that rewards go to venturesome and fortunate people rather than the analytical.[1]

In plain English, we speak about uncertainty as meaning that we don't know how events will unfold. And this being the case, it would be normal to say that risk (the possibility of one of a number of outcomes occurring) springs from uncertainty. It is because we know too little about the future that we cannot be sure that things will happen in only one way. Unfortunately, a technical interpretation is sometimes added to the terms risk and uncertainty. Risk is said to be present whenever there is more than one possible outcome to a process, and where we know what the probabilities of each conceivable outcome are. Uncertainty is present where we can conceive of different outcomes, but where we cannot estimate the probabilities that they will occur. This is a useful distinction, and the two different types of situation suggest different methods of analysis. The terminology, however, is not an obvious one and it will not be followed in the subsequent sections.

Risk and innovation

Risk is unquestionably a prominent feature of new technology. Innovation implies a certain desire to break from the traditional patterns of activity, so there tends to be less information available about its effects than there would otherwise be. But it is as well to remember that companies aren't closed systems – that their relationships with the outside world are central to their activities. Since the outside world is changing, uncertainty is unavoidable. Since innovation expresses a will to mould future events it may, in some circumstances, even be truer to say that an innovative course of action is one about which there will be the least uncertainty.

There are several types of risk that accompany new manufacturing technologies. Perhaps the most obvious are those associated directly with the technical properties of new systems. There is less historical experience on which to base estimates of development activities, so there are substantial risks of over-extending budgets and over-running delivery times. New techniques are malleable and have an intrinsic appeal to technologists, so there tends to be a greater risk of continuously enhancing the facilities a system provides beyond those that have been planned and those that can be fully exploited.

Close to technical frontiers there is the possibility of stepping beyond the state of the art, and of having to devote considerable resources to experimentation before working systems are developed. There is frequently a lack of information about how far proprietary products have to be modified to suit the conditions within specific companies: most people rightly expect computer-based systems to be adapted to their organization and its processes, and not the other way around. There is also a risk of obsolescence: of the technology making such rapid progress that technologists lose interest in manufacturing projects, and of software suppliers losing the depth of knowledge needed to support older products.

Another class of risk is less a product of the technology than of the way it is applied to a firm's operations. It is often hard to predict the disruption to business that will be experienced during a new system's installation – its call on resources (especially people's time) and the delays and errors it introduces temporarily to production processes. It can be hard to maintain a proper focus on the application, rather than the internal nature of a system, because it is the latter that is frequently more challenging and prestigious. To say that you have introduced a vision-guided robot has a distinction that talking about a more repeatable assembly process simply doesn't have. There are risks that the inability to measure the benefits of a new system will mean that there is too little incentive to achieve the effects intended. And, as we have already seen, the fact that information systems have a great reach across different departments amplifies the risk that all the relevant expertise isn't fully called upon during planning and installation.

Figure 8.1 Obvious sources of risk

When technologies are bought from other companies there are commercial risks. Contracts are invariably incomplete, in the sense that they cannot prescribe exhaustive courses of action for every contingency. In cases where it is difficult to foresee the nature of these contingencies it is obviously difficult to form contracts that both supplier and customer can be happy will be to their mutual advantage. There are risks that the contract will not be properly enforced – that too little effort will be spent in monitoring the progress of the work covered by the contract, or that the supplier will become unable to meet his obligations.

There is, as well, the danger of a diluted commitment within the factory to see a new system become successful. This is perhaps one of a third class of risks that follow from organizational issues. In what have been called mechanistic firms,[2] there is a tendency to skirt around major changes by creating new jobs or new departments in order to avoid disruption to the core of the existing system. It is existing structures and existing tasks that need to be overhauled. There is a risk that managers will react to the change in the firm's processes by looking for *more* mechanistic approaches – by spelling out procedures more definitively and by making job descriptions more rigid. This may well wreck the informal mechanisms that had earlier arisen to cope with over-rigid control. And advanced technologies characteristically serve to draw out the differences between people. Attitudes to new systems reveal who is innovative and who is traditional: who is visionary and who is blinkered; who is flexible and who is unbending.

Figure 8.1 provides a summary of some of the more obvious sources of risk.

Why non-diversifiable risk is not enough

Uncertainty about future actions has two main aspects – the real lack of information about conditions in the future (sometimes called the *objective* aspect), and the way in which ignorance about the future predisposes people to action (the *subjective* aspect).[3] Both aspects have their practical implications, and some of these will be explored later. But it is the subjective aspect that makes risk worth exploring in the first place. It is because the ideas discussed in Chapter 5 do not fully explain the way in which industrial managers react to risk that its treatment has to be taken further.

The argument made in Chapter 5 was that the one component of risk that is material to the value of a development is the *non-diversifiable* element. The net present value is found by applying a discount rate that is determined entirely by the opportunity costs of applying capital to that development. And it is only the extent of the risk that a firm's shareholders cannot remove by diversifying their financial assets that contributes towards these opportunity costs. In fact, when we calculate the discount rate we don't need to measure risk at all – we need only find the relative sensitivity of the project under appraisal to the factors which lead to risk.

However, this line of reasoning doesn't fully capture the price that the firm's employees exact for their part in accepting risk. As much as the firm's owners risk the returns on the funds they provide, its employees risk their pay-packets. In fact, employees are mostly faced with risks that are much more significant, personally, than those suffered by owners. In particular, the employees cannot diversify their contributions in the same way that shareholders can. They are more-or-less stuck with serving one employer at once, and the risks they take when they work for a particular firm stem from anything that might affect their earnings. Any mistake on the part of the firm's managers, such as bad marketing or production decisions, may imperil their jobs. As might any external factor which affects its operations – whether it is a factor restricted to the company's markets, or one that acts throughout the economy as a whole. Whatever might reduce their earnings, their expectations of better earnings, or threaten their jobs entirely, is a risk that is material to employees.

Since employees can't diversify, they are likely to take a much broader view of risk than shareholders. There are likely to be a good many more of the possible future states of the world that harm their interests. Some of these states will include redundancy, and perhaps the failure of the firm as a whole. Even in such extreme circumstances, a fully-diversified shareholder will lose relatively little, while an employee may lose a livelihood.

The upshot is that it is a wider concept of risk that, in reality, influences investment decisions. The company's employees, especially its managers, are most unlikely to be satisfied simply with acting on the level of non-diversifiable risk they perceive. This may from the financiers' point of view be unfortunate, and run counter to their interests; but it is inescapable, and any practical treatment of investment appraisal ought to acknowledge its significance.

This suggests that some developments with a positive net present value may well be rejected because they exceed the limits of risk that managers are prepared to countenance. It means that present value is an incomplete yardstick – that there is another dimension relevant to decision making. This does not, however, render present value redundant. One could, for instance, perfectly well retain a weak form of decision rule that made positive NPV (plus GOV) a permitting, but not a committing, test. In other words, all acceptable projects should have a positive NPV, but having a positive NPV doesn't guarantee acceptance. Such a rule is perhaps consistent with managers wanting the firm, above all, to *survive*. For

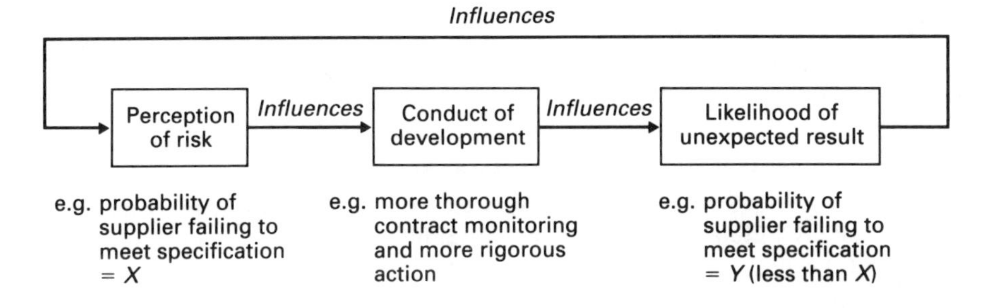

Figure 8.2 The influence of perceived risk on actual risk

accepting projects with a negative NPV reduces the worth of the firm and can only lessen its chances of survival.

This idea underlies the rest of this chapter. It simply means that we need to find the present value and option value in order to gain an understanding of a development's financial value, but that we shall apply in additional filter that reflects the attitudes of a firm's employees towards risk. There is, however, a danger that the process by which managers deal with risk is so intuitive that it is never properly examined, and never re-examined when major changes in the world at large suggest that it needs to be updated. We might be especially concerned that uncertainties about new technology are mis-interpreted: that the scale of these uncertainties is wrongly perceived or that the actions taken to deal with them are inappropriate. The bulk of the chapter is therefore concerned with understanding how people think about uncertainty, how it may be made explicit when firms take decisions about new developments, and what kinds of tool can be applied to give some consistency to its assessment.

The importance of volition

Many ideas about uncertainty assume that technologists and managers are faced with a number of alternatives whenever they make a decision, and that the outcome of each alternative course of action is entirely predetermined. This is of course rarely the case, because they can generally do a good many things to determine the success of their action after they have made the decision to take it. If a project begins to go wrong at some stage, few would put the turn of events down to fate, and few would let fate take its course. Most would probably attempt to correct it. The fact that people are identified with the decisions they make means generally that they are resolved to bring about a favourable outcome.

The way a development is carried out is, like the decision to embark upon it in the first place, influenced by the risks it is perceived to carry. If the risk of an FMS being defeated by commercial difficulties with the supplier is thought to be significant, greater attention will be paid to contract management. This means that the perceived risk has influenced the actual risk: one cannot say that action A has a probability of failing of X, because the fact that this is anticipated might suggest to people that steps should be taken to reduce this probability: to Y say (Figure 8.2). A more general way of describing this effect is to say that we cannot separate what we observe from our act of observing it.

The idea that technologists and managers exercise large amounts of free will is countered to some degree by the frequent observation that their decisions are made for them by changes in the environment, and by the apparent existence of sub-conscious thresholds of tolerance that only now and then trigger them into doing something. It is a frequent admission that one's conduct is wholly driven by events. It is therefore probably reasonable to talk about most projects having one most likely outcome, and a certain probability of going wrong. But such an assessment should not be interpreted as describing something that is entirely beyond the influence of a firm's employees.

The problem of bias

Unfortunately there are not only random uncertainties in the predictions people make, but also various kinds of consistent error. There are, for example, a number of biases which people are known to apply whenever they make judgements on the basis of experience or rules of thumb.[4] One of these is a tendency to pay too much attention to the most readily available information. The results of this are typically that people predict the events that they can most easily imagine taking place, or incorporate considerations that they can most readily recall from past experiences. A second short cut is that they often apply illusory correlations, overestimating the extent to which two influences are associated with one another. A third is that their thoughts tend to become anchored to a particular figure or issue. Even when new information becomes available, they fail to revise their estimates sufficiently to reflect this new information.

Such effects have a particular impact when we are trying to make predictions across a substantial step-change in technology – when we are trying to estimate the nature and the scale of effects in a world based on a number of principles that differ from those applicable to present conditions. Paying too much attention to readily available information suggests that people will naturally adopt the assumptions of their own information systems – their firm's management accounts – although the limits to their applicability are widely understood. Seeing illusory correlations is typically the reason that so much stress is laid on utilization, in the expectation that machines and human beings in continuous motion are inevitably associated with worthwhile economic activity. And the problem of anchoring is evident in the belief that customers focus primarily on product prices, when there is information available that indicates that other product qualities are noticeably more important. This sometimes leads to an unchanging pre-occupation with reducing costs, when it can be demonstrated that the more significant and more probable benefits of a technology lie elsewhere.

Problems also arise with collective forecasts. You do not have to be a complete sycophant to tend to concur with your superiors, and groups of people often attempt to preserve their cohesion by assuming that they are invulnerable to mistakes. It is not unknown for groups of quite senior people to isolate themselves from the day-to-day activities whose conduct they direct.

A recent survey of the feelings of finance officers in large American companies[5] showed that some four-fifths of them thought that revenues were typically overstated in capital budgeting proposals. Of this 80%, around a third attributed the bias to intentional manipulation, and another third to inexperience on the part of their analysts. The final third thought the bias came from two main sources. The first was the influence of group polarization effects, in which the positive opinions of individuals are magnified in a group of people as a whole. The result is that the benefits of a project are often overstated and the likely difficulties underestimated. The second source of bias originated in the poor quality of information supplied by senior managers, who perhaps had pet projects they wanted to see treated favourably. That they could get away with this was seen as an indication that bias arises wherever a company operates inappropriate reward structures – structures that don't induce people to carry out actions favourable to the company as a whole. As you might expect, the survey also found that when investment proposals called for a significant departure from established processes, the finance officers expected bigger forecasting errors.

Feelings about bias in cost estimates were similar to those about revenue estimates. On average, the people questioned claimed that the actual costs incurred in new projects were 15% higher than those predicted during investment appraisal. Perhaps needless to say, outcomes were more likely to differ from forecasts when proposals incorporated significant advances in production technology.

The picture painted by the survey is rather a bleak one, for it suggests that refined yardsticks are being applied to coarse data. Of course it is important to bear in mind that it represents the individual judgements of finance officers: it is more a survey of perceptions than one of measurable events. But it is obvious that less bias will be good for the appraisal process. It will reduce the uncertainty in the returns to a firm's developments and give people confidence that an evaluation is a satisfactory basis for making decisions. The fact that just 50% of the people responding to the survey had confidence in the appraisal process indicates that firms could do considerably more than they do at the moment, on average, to tackle bias.

They might, to begin with, consider improving the stock of knowledge and experience on which the people conducting evaluations work – whether by training, recruiting or calling on the services of outsiders. They also need to eliminate as far as possible the incentives for overstating profitability: perhaps by making the appraisal task a more prominent element in a person's job, so that the accuracy of his forecasts will contribute towards his salary rises (or the lack of them).

It seems quite likely, in fact, that the rather extended nature of many firms' decision-making processes reflects the need to test proposals for bias by exposing them as widely, and for as long, as possible before they are acted upon. This does not help offset the biases characteristic of people working in groups. But the ways of tackling group biases are also rather obvious: leaders shouldn't offer their views before others have offered theirs, and it is helpful to have a devil's advocate to avoid the illusion of collective invulnerability.

In contrast to the problem of random uncertainty, consistent bias ought to be more straightforward to recognize. By definition it acts in a consistent direction, and is open to consistent remedy or compensation. Tackling bias is perhaps less a case of solving the technical problems of identifying it, and more a case of facing up to the managerial problem of eliminating its origins.

Action

1: Invest in IGES
2: Do nothing

State

1: Customer provides much business, products remain electro-mechanical
2: Customer provides little business, products remain electro-mechanical
3: Customer provides much business, products become electronic
4: Customer provides little business, products become electronic

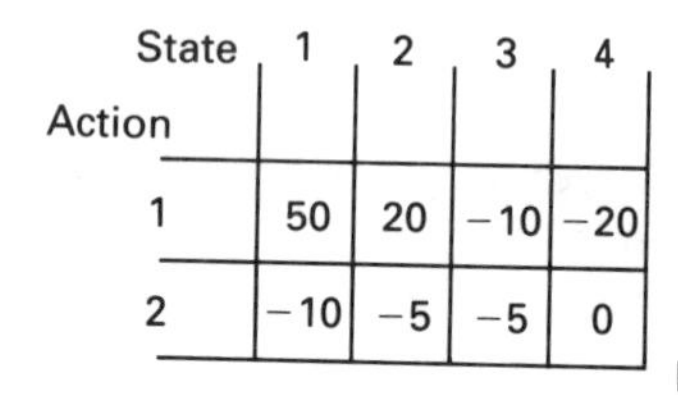

Action \ State	1	2	3	4
1	50	20	−10	−20
2	−10	−5	−5	0

Payoff matrix:
NPV in £000s

Figure 8.3 Payoffs in different states of the world

8.2 Payoff strategies

Actions and states

We can now look at various ways of analysing risky or uncertain situations in detail.

To carry out the most basic type of analysis, we need to be able to identify a set of future *states* which are beyond a decision maker's control, and a set of *actions* which are within his control. The states describe properties of the world at large that affect the worth of an investment, and each state is a distinct combination of these properties. A state might encapsulate the strength of the national economy and the condition of a particular product market, perhaps. The actions can be initiated at the discretion of the person or group making investment decisions. In the simplest cases there may be just two actions: one to proceed with a new development. and one to terminate it.

The method of analysis is to identify, for each combination of action and state, how favourable the outcome is likely to be. When the actions refer to investments in technology, these outcomes will probably be stated in terms of net present value and growth option value. Any one of a number of strategies can then be applied to select the best of the available actions.

Suppose, for instance, that a company is considering whether to buy IGES pre- and post-processors so that it can exchange product drawings electronically with a specific customer. The value of the work, or its payoff, depends on a number of circumstances in the outside world: whether the customer continues to be a customer, whether he is likely to want electronic drawing exchanges, whether he adopts IGES translators himself, whether IGES will continue to suffice for product designs, and so on. Equally, the payoff of *forgoing* the investment in IGES will depend on these same circumstances. It may be risky not to proceed if the customer's opinion of its suppliers is coloured by their unwillingness to invest in such a technology.

The various payoffs that are obtained in particular states, given that particular actions are taken, can be summarized in a matrix such as the one shown in Figure 8.3.

For the sake of argument, it has been assumed that there are just two properties of the world's state that are material to the worth of the IGES investment: the volume of business done with the customer in question, and the nature of the products that will in future be sold to him. Since each property takes one of only two values, there are in fact four possible states.

From the figure, it is fairly plain that in states in which IGES proves to be a great benefit, the action of not investing has a poor payoff — and vice versa. It is also evident that neither option is *dominant* in the sense that neither is invariably better than the other, no matter how the future turns out. There are some states in which the first option would have the better payoff, some in which the second would. To make a decision about which is the preferable option we therefore need to take a closer look at the pattern of payoffs.

Decisions under uncertainty

One approach to making a decision on the basis of this information would be to take the action that minimizes the maximum possible loss (that is, the loss in the worst state of the world). This captures a highly conservative approach to dealing with unceraintry. It obviously implies that IGES should *not* be pursued, since the worst outcome is in State 4, following Action 1. (It appears as though the cost of the project is £20000, because the benefit of IGES in State 4 is likely to be zero, and the payoff is −£20000). Such an approach is pessimistic because it ignores all but the most unfavourable future states. An obvious alternative is to maximize the maximum gain, and, in this case, the decision would fall in favour of IGES since it promises a payoff of £50000 in State 1.

A more sophisticated approach is to minimize the maximum possible *regret*, where the regret is defined as being the difference in any particular state between a given payoff and the best possible payoff in that same State. For example, in State 2, the regret associated with Action 1 is zero, since Action 1 gives the best possible outcome. Action 2, however, has a regret of £25000 because if State 2 occurs the firm could have made this much more by taking a different action. This strategy, known as minimax regret, can again be illustrated with matrices showing the payoffs, the regrets derived from them, and then the worst regret associated with each action (Figure 8.4). In this case, Action 1 is preferred because it minimizes the worst regret to £20000.

There are several other strategies of this type, including some that generalize on the ones just described. Details may be found in many books about, or with sections on, the subject of uncertainty[6].

Real situations are rarely so simple that there are just two available actions and four possible states of the world. Even with such a small project as our IGES scheme there might be a third action that involves sharing the cost of development with the customer in question — and perhaps some contractual agreement on a minimum level of future business. This would be a kind of hedge on how the future

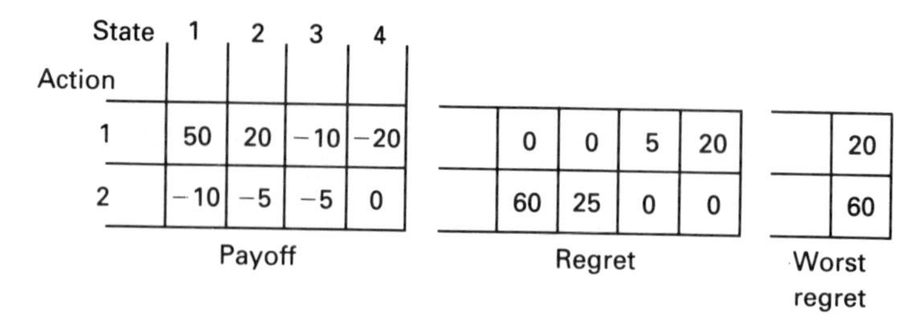

Figure 8.4 Working out regrets from payoffs

turns out: it is a half-way point between two actions that are thought to be optimal in different future states of the world.

And it may be the case that we have underestimated the number of future states. These states are only distinguished by their essential properties: they lump together every possible state that has a particular set of values for these characteristics. So, in our example, State 1 includes every possible pattern of future events in which our customer continues to provide business and in which the products remain electro-mechanical in nature. But in practice it often turns out that there is a variable which affects the course of a development, but which isn't considered in advance. We might have forgotten that there is a possibility that the customer will buy a CAD system that is currently not provided with an IGES post-processor.

This suggests that there are often grounds for being more uncertain than one in fact is. It is quite possible to overestimate the amount of information at hand.

8.3 Expected values

Subjective probabilities

When we have a feeling for the likelihood of uncontrollable events turning out in a certain way, it makes sense to incorporate this additional information in our approach to analysing risk. The traditional method of doing this is to look upon the likelihood of events and the impact of their occurrence in actuarial ways. These are simply based on predicting the likelihood of future events by assigning them probabilities — numbers that denote the frequencies with which certain things happen if the same circumstances are repeated on a large number of occasions. If a coin is thrown a hundred times, and found to land on a particular side in forty-nine of them, you can then say that in the absence of further information there is a probability of 0.49 that the next time the coin is thrown it will again land on that same side.

Although the process of making industrial decisions is far more complex than throwing coins, we hope, the same sort of reasoning can be applied. Perhaps it can be estimated that, forty-nine times out of a hundred, a development of similar characteristics to the one under evaluation would make a loss. In other words, there is a 49% probability that a state occurs in which the project loses money. Here, however, the source of information about this risk is clearly different from that in the coin-throwing case: whereas that was based on repeating an action many times to gain an *objective* understanding of the coin's behaviour, it is impossible to conduct a project a hundred times. The estimate of the probability of the project making a loss is therefore a *subjective* one.

It is not enough, in any case, to know only about the probability of an outcome. If it is to be set in proper proportion with all the other possible outcomes, some measure of its effects will also be needed. To say that in a particular outcome a project makes a loss is really to say very little, for instance. It would be much more useful to know in how many of the hypothetical forty-nine times out of a hundred there is a possibility of bankrupting the firm. If it were one out of forty-nine then the project might be considered an acceptable risk, while if it were forty out of forty-nine it would almost certainly not be.

Expected value

The most common way of combining the notions of probability and impact is to form expected values. If there were a 20% probability of making a loss of £1M then you could speak of an expected loss from the associated turn of events of (20% × £1M), or £200000. If the same course of action had an 80% probability of making a gain of £800000, then the action as a whole has an expected value of (80% × £800000) – (£20% × £1M), or £440000.

In fact, when we speak of expected values we are not always expressing how worthwhile we perceive a course of action to be. This is because there is probably no simple, linear relationship between expected monetary value and personal reward or satisfaction. For example, the fact that a thousand pounds is a lesser reward for a rich person than a poor one indicates that there is a diminishing, marginal satisfaction to monetary value. In principle, we ought therefore to look for a measure of what we might call utility, and translate our expected values into expected utilities. In practice, it is impossible to measure properly a person's utility function, and such notions are too individualistic for commercial decision making. Commercial firms do not embody, after all, a single person's values. We shall therefore continue to consider expected values, while acknowledging that a million pounds is more appealing to a near-insolvent firm than it is to one that is prospering.

Suppose in the IGES example used in the previous section we had been able to assign probabilities to each of the four future states. We might have predicted, for instance, that there was a 30% chance both that the customer continued to provide good business, and that the products would remain electro-mechanical. In other words, that State 1 had a probability of 30%. We can add this, and the likelihoods of the other states, to Figure 8.3 and reproduce this as Figure 8.5. (Since there cannot be a future represented by anything other than one of the four states, their probabilities must add up to one.)

The expected value of each action is calculated by weighting the outcome of that action in each state by the probability of the state occurring, and adding these

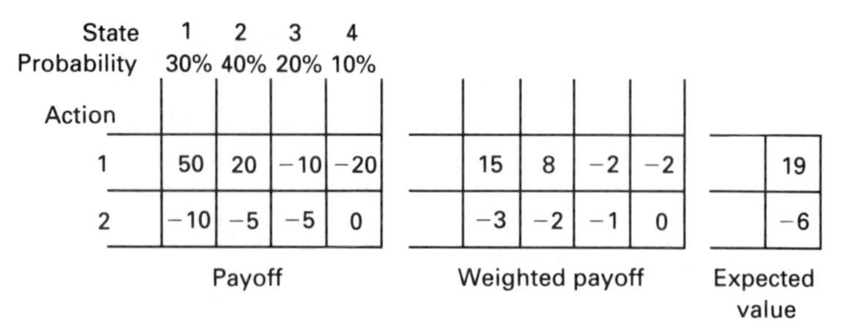

Figure 8.5 Payoffs, states and probabilities

weighted outcomes up. For instance, the expected value of Action 1, an investment in IGES is

$$£50000 \times 30\% + £20000 \times 40\% - £10000 \times 20\% - £20000 \times 10\%$$

or £19000. Similarly, the expected value of the alternative action is −£6000. So, provided we wanted to maximize expected value, we would clearly choose Action 1 and proceed with the IGES development. It is possible, in fact, to change the selection rule in the manner of the previous section where we considered *regret* in place of payoff. With expected values, however, minimizing regret always leads to the same choice of action as maximizing pay-offs, so there is little point in considering both.

For what it's worth, we can also perform another interesting manipulation now that we have a set of probabilities. If we knew in advance that State 1 was to occur, then we would select Action 1 because that yields the best result. Similarly, if we knew it was to be State 2, then we would also select Action 1. But if we either knew it was to be State 3, or knew it was to be State 4, then we would select Action 2. Therefore, it appears from our present condition of ignorance about the future that *if* we knew for sure what would occur, the expected pay-off would be

$$£50000 \times 30\% + £20000 \times 40\% - £5000 \times 20\% - 0 \times 10\%$$

or £22000. This is some £3000 better than the expected value we are predicting by taking Action 1. This difference is known as the expected value of perfect information, because it tells us the extra pay-off that we could expect from knowing which state represented the future. It therefore tells us the maximum amount of money we should consider spending in order to determine this.

Variance and risk

A problem with the analysis illustrated in Figure 8.5 is that the probabilities are based on the idea of repeatable events, while in reality large developments are anything but repeatable. This suggests that the decision maker cannot be concerned solely with the expected value of a particular development. Because he is making

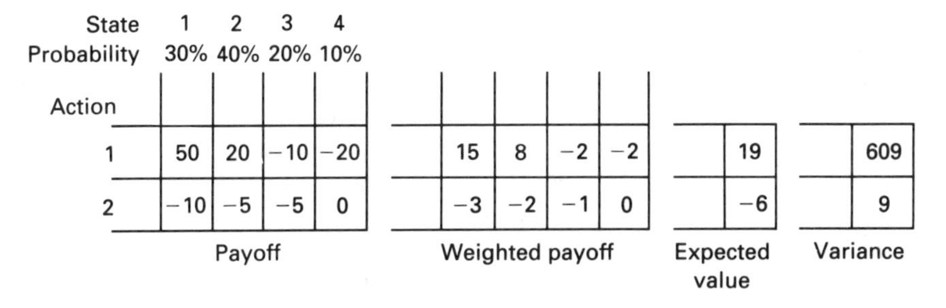

Figure 8.6 Actions and variances

a decision about carrying out a project just once, he cannot hope that in the long run its outcome will approach the average: if the outcome turns out badly, that is the end of it. He will therefore be concerned with how dispersed the outcomes of his actions are in different future states. All other things being equal, he is more likely to choose a project that would have the same outcome whatever the nature of the future than one whose outcome is heavily dependent on uncontrollable events turning out favourably.

The common way of measuring this kind of dispersion is to take a variance. The variance is simply the mean of the squared deviations of each outcome from the expected value. Continuing with our example from Figure 8.5, reproduced in Figure 8.6. we can find the variance of Action 1 as

$$30\% \times (50 - 19)^2 + 40\% \times (20 - 19)^2 + 20\% \times (-10 - 19)^2 + 10\% \times (-20 - 19)^2 = 609$$

The variance of Action 2 is substantially less, at just 9. The difference in the dispersion of outcomes in the two cases can of course be seen without going to the trouble of calculating the variance: it is obvious that in Action 1 the outcome will either be very good or very bad, while, in Action 2, it will only be slightly good or slightly bad. In other cases, the action with the lesser dispersion may not be evident simply by inspection.

We could argue that a development which shows a wide dispersion of possible outcomes is more risky, since there is a greater likelihood that the results will deviate appreciably from the expected value. In other words, we can regard variance as an index of risk. (In practice, the potential for under-performing seems to be more important to industrial managers than the possibility of over-performing, and there is evidence that the decision-making process focuses more on semi-variance than variance.[7] This is a measure of the half-width of a probability distribution, up to but not beyond the expected value.)

We now have two quantities influencing decisions on whether to pursue a project: the expected value (or mean) and the variance – standing, respectively, for return and risk. This means that we again have problems when comparing two, alternative courses of action if neither is dominant. In other words, when one action has the greater mean and the lesser variance it is definitely the better action; but when one has a greater mean, and the other the lesser variance, we do not so far have a way of choosing among them. This is plainly the case in Figure 8.6, for while Action 1 has a favourable mean it has an unfavourable variance. We therefore want to be able to reduce the two dimensions to a single scale: to trade off variance against mean, or vice versa. Much of the theory about financial markets (including the derivation of the capital asset pricing model mentioned in Chapter 5) is based on the assumption that people do make such a trade-off in practice.

In markets, inferences can be drawn about this trade-off by observing the prices and characteristics of the commodities that are exchanged. But within a company it is a harder thing to really know, even for one person, what the trade-off ratio should be. It will probably vary from day to day, and it will not be constant over all values of risk. And, in real firms, investment decisions tend not to be made by one individual, although one individual commonly has the final sanction. This makes finding a suitable trade-off trickier still.

There are various devices that can be used to avoid confronting this problem – notably by looking at technical properties of the probability distribution[8]. But if our analysis is intended to reflect the way a decision maker would react to risk intuitively, given all the relevant information, we need to be wary of using sophisticated mathematical techniques which couldn't possibly be a basis for this intuitive reasoning. Perhaps the most appropriate way of working is to present decision makers with predictions of mean and variance, and tell them to draw their own conclusions as to whether the size of the mean justifies that of the variance. Doing things this way at least restricts the scope of the judgement that is called for.

Qualitative analysis

The final type of analysis I want to mention while we are still considering expected values is one in which the analyst has an imprecise idea of the probabilities of future states. He has some feel for their relative likelihoods, but he cannot quantify them. It is a condition mid-way between that of Section 8.2 (in which no information about probabilities was available) and that of the earlier parts of this section (in which there was a perfect knowledge of these probabilities).

Qualitative reasoning is the term often used to describe a line of argument in which one can speak about elements on an ordinal scale but not a cardinal scale. In other words, it might be possible to say that one state is more likely than another, but not to say how great the difference in their likelihoods is. This is therefore an example of a situation in which there is an inexact understanding of probabilities. Because it seems to be such a realistic situation, it would be nice to find a model that would help us reach decisions in such conditions. Unfortunately, rigorous reasoning using qualitative models of probability is relatively hard, and in practice the calculations that one could perform are pretty intractable. In the remainder of this section I shall outline an approach due to Buehler[9] in an attempt just to convey the flavour of a qualitative model.

We begin by assuming that the analyst can identify all future states of the world, and all possible actions, together with the outcome of each combination of action and state (as in Section 8.2). In addition, he can at least partially say which states are more likely than others. He might, for instance, consider that State 1 is more probable than State 2. He might, at the same time, be quite unable to say how the likelihoods of these states compare with that of State 3. Suppose in fact that in the IGES example of Section 8.2 we could make the following statements of relative probability; the symbol '$\geqslant$' means 'no less likely than'):

State 1 $\geqslant$ State 2; State 3 $\geqslant$ 4

If you consult Figure 8.3 you will see that this statement suggests that we know it is more likely than not that the customer will continue to provide a good level of business. We could only compare the likelihoods of, say, States 2 and 4 if we were able to say whether it was more or less likely that the products would become electronic. In this case we evidently can't do this.

There are, of course, a great many probability distributions that are consistent with the partial ordering that we have just obtained. For example, if States 1 to 4 had probabilities of 0.4, 0.3, 0.2 and 0.1 (respectively) then this would be consistent with our purely qualitative knowledge. In general we can speak about a set P of single probability distributions p that are consistent with the partial ordering. Each member of P is a short list containing, in order, the probabilities of each state, such that they add up to one. A member of P is obviously $\langle 0.4, 0.3, 0.2, 0.1 \rangle$. Our decision-making procedure will be to select a suitable member of P and then work out the expected value of each action as though this p were in fact the correct probability distribution (much as in the earliest part of this section). The key is obviously to find a satisfactory p.

The fundamental rule that Buehler uses is that the optimal course of action is the one that achieves the best results against the background of the worst conceivable turn of events. This means that the decision-making procedure is, first, to find the probability distribution p, that, being consistent with the partial qualitative knowledge, gives us the most pessimistic view of each action's expected value. And, second, to choose the action which maximizes this expected value.

The way in which this is done is unfortunately too involved to reproduce here. However it should be apparent that the essence of qualitative reasoning is to identify a partial ordering of some sort, and then choose a number of cardinal values that are both compatible with this ordering and satisfy some other criterion. The problem is then solved as though these cardinal values represented our true knowledge about the situation. The procedure is inevitably complex, and perhaps too extended to be applicable in normal industrial decision-making processes.

8.4 Focus outcomes

Actuarial drawbacks

Although they are superficially attractive, the actuarial ideas of probability and expected value begin to look less applicable when industrial investments are considered in more detail. For a start, we have already seen that these investments are not generally repeatable. Section 8.3 began by speaking of risk in terms of the probabilities found by repeatedly taking the same action and recording the results: if something happened 49 times out of a hundred you could speak of a probability of 0.49. It was assumed, tacitly, that if a new development was implemented one hundred times then 49 of these times it would make a loss. This is of course a nonsense, since even if a firm had the luxury of doing the same thing a hundred times it would accumulate new information and expertise at every go. As a result, it would never perform the project identically from one occasion to the next. When we assign probabilities to industrial projects we cannot really be speaking about the frequencies of specific events occurring in response to repeated processes.

Industrial investments are also never very divisible: it is hardly possible to carry out 33% of one development, and 33% of each of its two alternatives. So, in calculating expected values, the analyst is effectively taking the average of outcomes that are mutually exclusive – outcomes that are rivals to all intents and purposes.[10] This can sometimes seem highly paradoxical. If for example, a project had a 50% probability of earning £150000 and a 50% probability of earning £50000, it has an expected value of £100000 – even though it could never turn out to earn this value (only £150000 or £50000). To say that you expect an outcome which you claim to be impossible would hardly enhance your credibility as a decision maker.

Assigning probabilities also involves applying a *cardinal* measure of uncertainty: an absolute scale, running from 0 to 1, against which the likelihood of an outcome occurring is placed. This implies that there can be an indefinitely fine *ordinal* scale of probabilities – that you can always say which of two outcomes is more likely than another unless they are unequivocally identical. This, again, is entirely unrealistic when it is complex and uncertain developments such as computer-based systems that are at issue. For there are always large margins of error associated with our estimates of how likely a future event is. This second-degree uncertainty makes reasoning about probabilities a still more unsatisfactory way of understanding risk. It suggests that we are vulnerable to the dangers both of misunderstanding the influence that risk has on our actions, and of thinking that our calculations have a precision which they completely lack.

Potential surprise

There are some significantly different ideas about how people actually look and act upon uncertainty. It is worth briefly examining these because at the very least they serve to illustrate the kind of information that decision makers are likely to receive when they ask, say, technology experts about their expectations for a particular development.

It has become evident that probabilities as frequencies are not a good way of understanding people's beliefs and doubts about the future of a new development – beliefs and doubts that are essentially products of the mind rather than dispassionate observations. A way of looking at these products of the mind is based on the concept that people think about uncertainty in terms of surprise: the surprise that they would potentially feel as a result of the occurrence or non-occurrence of a particular outcome to a particular course of action.[11] The potential surprise associated with a decision is *not* something found by the sort of coin-throwing experiment that characterizes probabilities. It is a subjective notion that reflects the way a person forms expectations about the future.

Curves of this potential surprise can be drawn against specific scales of characteristics that tend towards increasingly bad values in going to one extreme and increasingly good towards the other. An obvious scale to use here is that of financial value, since we are most interested in a technologist's or manager's view of how a development's worth might differ from its anticipated value. We would probably use net present value, together with growth option value if it can be quantified.

For example, the NPV of a project might be estimated at around £100000. At the same time the experts proposing this figure could argue that they would in fact be unsurprised if the measured worth, once the project was in place, turned out to be somewhere between £80000 and £125000. On the other hand, they would be astonished if it were less than £50000, or greater than £140000. This can be expressed pictorially in the manner of Figure 8.7. The broken line represents the relationship between worth and potential surprise in the perception of our experts.

Such curves of potential surprise are seldom perfectly static. General experience, for example, tends to decrease surprise; older men are, apparently, unsurprised at wide departures from calculated outcomes, unlike youngsters who frequently act on their calculations as though they were certainties.[12] The curves can also shift suddenly when something that was novel and previously unexperienced becomes the state of the art and is in turn ready to be superseded by something newer. It is natural to suppose, for instance, that different curves would apply to a manager's view of mechanized technology and his view of automation. His surprise curves for automation might be a good deal broader than those for mechanization, indicating that it would not really astonish him if the economic outcome of automation were either very favourable indeed, or not favourable at all.

It is sometimes difficult to see the distinction between potential surprise and probability, and an example can make it more obvious. If there were an action that had N possible outcomes, all equally likely, then the probability of each would be $1/N$, since the probabilities of all the outcomes must add up to one. If the action were that of rolling a die, N would be 6. The probability of rolling a two, say, would be one-sixth – which is quite small. This does not, however mean that you

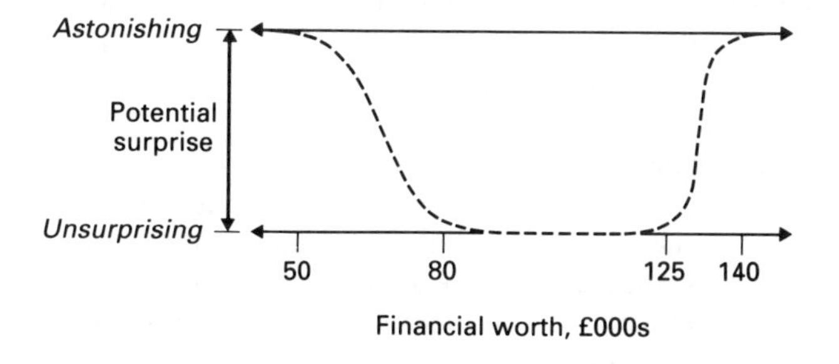

Figure 8.7 Curves of potential surprise

would be at all surprised about rolling a two. If the die had a thousand sides, each marked differently, the probability of rolling a two would be very slight indeed: and yet the potential surprise at rolling a two would be unchanged. It has to land on one side, after all, and a two is as good as any other.

The theory of potential surprise is unproven as an accurate description of the way in which people behave in the presence of uncertainty. There is no guarantee that an expert's surprise profile, even if it can be demonstrated that he really experiences it, is an appropriate one: he might be ignorant or prejudiced about the subject of his surprise. But the idea of potential surprise does give us an interesting framework for bringing expectations out into the open.

Focus outcomes

As well as associating a degree of potential surprise with a specific outcome, it is normal for someone to feel a pleasure or distress in anticipation of that outcome. Pleasure will typically be felt when anticipating profits, and distress when anticipating losses. Since we are speaking about a subjective judgement of an action's impact, about a product of the mind, we have to understand that we can only record present feelings about future events. We cannot claim to record the impact on the decision maker's mind of those events when they take place, since that lies at some point in the future. Generally, however, it is reasonable to say that the more extreme the anticipated economic value of a development, the more intensely a firm's decision makers would feel when they anticipated it.

There will, however, be a point where the degree of potential surprise reaches such a level that the impact of any outcome associated with an even greater surprise would be ignored, or at least discounted. Although an impact may seem very high, its chances might appear to be too remote to be worth paying much attention to. The critical point at which surprise and impact combine to have the greatest effect is where the attention of a decision maker is thought to be most closely focused. The outcome represented by this point is known as a focus outcome.

In Figure 8.8, the first sketch is of a potential surprise curve very similar to that of Figure 8.7. The second sketch combines the potential surprise with the idea of impact in anticipation to form a function that tells us, in the mind of the person who provides this information, how significant each possible outcome is. In particular, there are two local maxima: these lie at the focus outcomes. The left-hand maximum is known as a focus loss, and it represents the point at which an increasingly unfavourable outcome is about to be outweighed by the increasing surprise that would attend its occurrence. The right-hand maximum is the focus gain.

The importance of focus outcomes stems from the observation that the attention of people making decisions tends to rest on the greatest, readily conceivable distress or pleasure they face – the outside possibilities as it were. The virtue of a particular course of action is then measured by weighing the two outside possibilities, one highly favourable and one highly unfavourable, against each other.[13] The manner in which this is done is suggested by the curves of indifference sketched in Figure 8.9. A decision maker whose behaviour is represented by these curves is indifferent

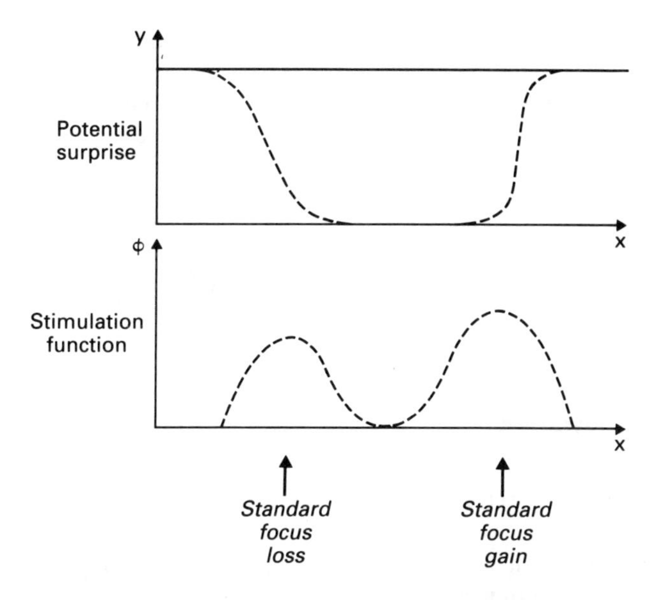

Figure 8.8 Focus outcomes

to the choice of points on a given curve: he will perceive the combination of loss and gain at each point on a single curve to be associated with actions of equal value. A curve that lies to the left of another represents actions that are more valuable, and a curve to the right represents actions that are less valuable. This trade-off of focus loss and gain is supposed to be evident in such common phrases of commercial life as 'not much downside risk and much upside potential'.[14]

Such an idea is in marked contrast to those approaches that lay stress on expected values – on outcomes close to the centre of a decision maker's anticipations. When you consider how much of their time people spend dreaming of highly favourable outcomes, and worrying about highly unfavourable ones, it seems quite plausible that they play a bigger part in decisions than do averages. The most extreme inference that could be drawn is that any approach to risk

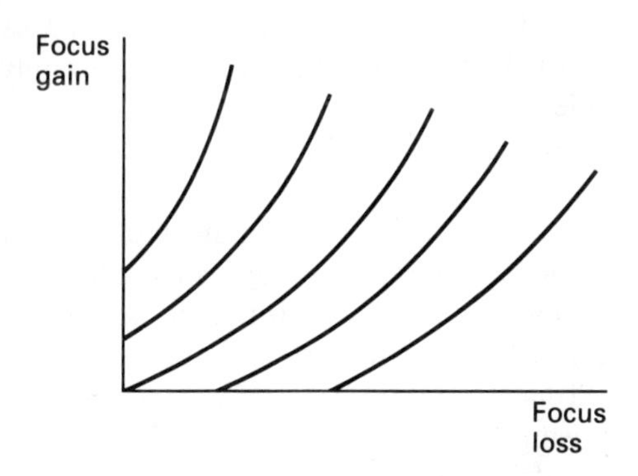

Figure 8.9 Indifference to focus loss and gain

analysis that yields only an expected value is worse than useless: it is costly to compile, yet it will be ignored when decisions are taken.

The idea of potential surprise was suggested as an explanation of the way people naturally form expectations and act as a consequence of their expectations. It is not, necessarily, a suitable tool of analysis in the eyes of managers intent on their companies pursuing a narrow range of simplified, measurable goals. But the descriptive nature of these ideas suggests that they provide a better way of expressing notions of uncertainty than the actuarial principles discussed earlier. As a result they might lead to a method of explicit analysis on which decision makers would be happier to act.

8.5 Practical choices

The choice of an appropriate way of understanding and dealing with risk is very much one for individuals, or at least individual firms, to make. We are, after all, attempting to acknowledge the fact that managers act according to risks that are material to themselves, not just risks that register on the net present value yardstick. To settle on one or two specific methods is essentially to make explicit a decision maker's natural feelings about how risk arises and how it should be controlled. It would be wrong to suggest that such a practical issue should be addressed in the same way in all firms and in all industries. The test of an effective way of managing risk is whether its users find its assumptions and mechanisms credible and useful, and whether it consistently delivers results that, over time, prove to be good indicators to important issues. While certain logical problems and implications can be pointed out – in the manner of the preceding sections – the proof lies in the application.

Identifying risk

The first step that normally needs to be taken is to understand where risk arises: to identify the matters about which people are sufficiently uncertain that there is an appreciable likelihood that they will cause a development to go awry.

One approach is to start by dividing a development into a series of activities, and then to look at their distinctive parameters. We might apply the idea of cost drivers, for instance, to identify the factors that have the greatest influence on the cost of performing the activities. Any uncertainty in these factors will lead to a risk that costs will over-run.

Suppose that we are considering the development of an expert system. This would probably involve the activities of buying a shell program from a software supplier, buying a computer from a distributor, eliciting an expert's knowledge, and gradually developing a knowledge base. The main costs are obviously the purchase of software and hardware, the time of the expert and analyst, and the time of managers supervising the progress of the work.

The purchase costs are driven by such elements as the amount of knowledge the system is expected to record, the number of people who will be able to use

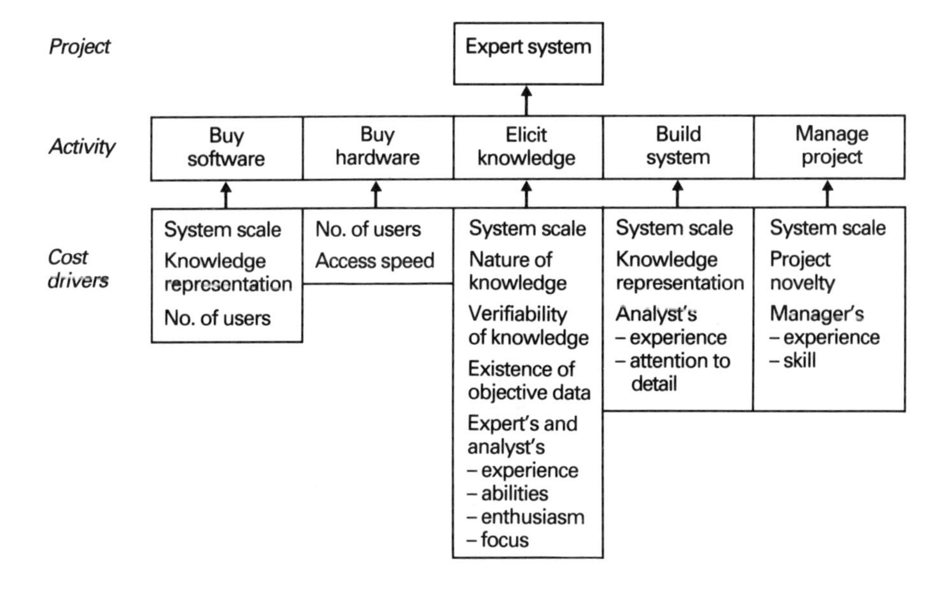

Figure 8.10 Cost influences for an expert system

the system simultaneously, the speed at which it must operate, and the nature of the knowledge that is to be captured. The amount of time the employees will spend on the work is driven by their experience, abilities, enthusiasm, their ability to focus on this one project and, again, by the scale of the system. It is also influenced by how readily the expert supplying the knowledge can articulate the way in which he characteristically solves problems, whether his reasoning can be deduced from specific examples, and whether there is any objective data that might confirm his claims. The breadth of uncertainty in the nature of these cost drivers – uncertainty about the team's enthusiasm, for example, and uncertainty about the sensitivity of costs to this enthusiasm – means risk (Figure 8.10).

In most cases, it will *not* be possible to apply a similar analysis to a development's benefits because they are rarely a function of the separate activities that take place during its introduction. The benefits are usually determined by the project's outputs – by their relevance, accuracy, timeliness and so forth. So, in this case, we might start simply by listing the assumptions that have been made during the prediction of a proposed project's performance. The small example discussed earlier, in which a firm is considering whether to adopt IGES, rested on assumptions about the nature of its products in the near-future, and on the customer continuing to provide a constant level of business.

The risk a firm faces of not achieving the predicted benefits is a function of the confidence with which these assumptions are made. The second step will therefore be to understand the degree of confidence attached to these assumptions. Probabilities might be used here, and if we felt that we knew a great deal about the relevant patterns of uncertainty we might even be able to attach complete probability distributions to those factors that have been quantified. It ought to be apparent by now, however, that these probabilities are flawed – that they are

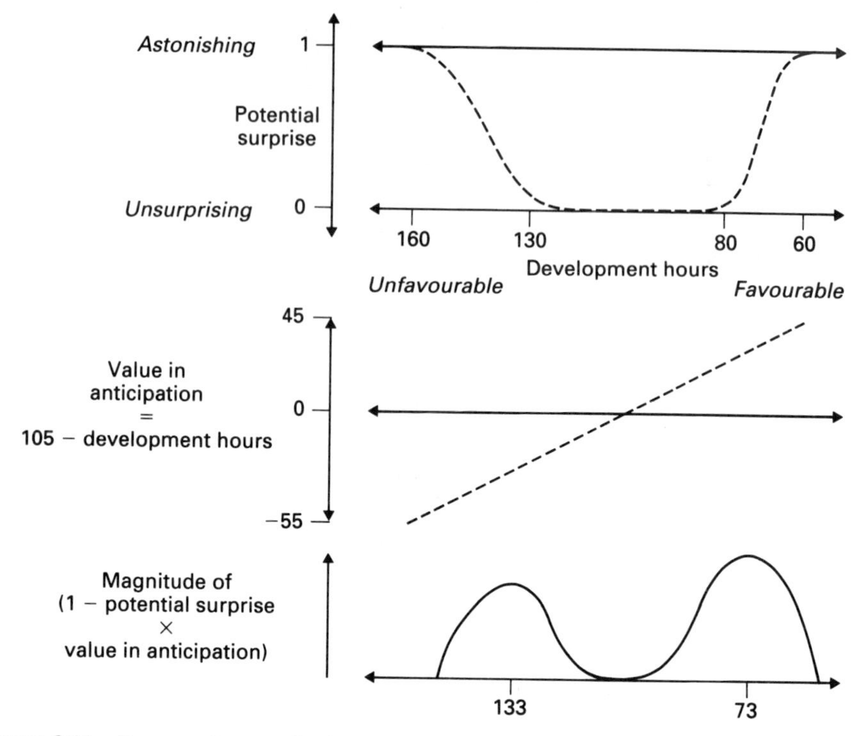

Figure 8.11 Focus outcomes in development hours

hard to measure and that they do not in any case seem to be a particularly good expression of how people form expectations about the future. An alternative is therefore to use the idea of potential surprise, and to sketch curves of the kind shown in Figure 8.8 for each of the factors that have a bearing on a project's outcome. We could, in our IGES example, say that it would be unsurprising if the project consumed between 80 and 130 development hours: but that it would be astonishing if it consumed less than 60 or more than 160.

The third step is to understand the impact of any lack of confidence in these determining factors. Although the principle of expected value is initially promising, there are again problems in applying it to industrial developments. This suggests that a more promising approach is to build on the potential surprise curves that might have been compiled on the previous step in the manner of Section 8.4. There the potential surprise was combined with the value of an outcome in anticipation in order to find focus losses and gains. It is not in fact obvious how we would do this explicitly, but one way is to assume that each of the determining factors lies on a scale that goes from bad at one extreme to good at the other, and that the value in anticipation is a linear function of this scale.

For instance, in Figure 8.11 there is a potential surprise curve that records the expectations of IGES development hours just mentioned. Below this is a line that *might* represent the value in anticipation: it assumes that the greater the number of development hours consumed by the project, the less pleasant the outcome appears. The line crosses the horizontal axis when the number of development hours is 105: this is the centre of our expectations. The final sketch is based on the assumption that the significance of an outcome is determined by the product of potential surprise and value in anticipation. The potential surprise has been re-expressed such that the absence of surprise is scored as one, and complete surprise is scored as zero. By taking the magnitude of this product, the points of focus loss and gain (at about 133 hours and 73 hours, respectively) are evident. This means that a manager's perception of the acceptability of the project will be determined by whether he thinks that the possibility of the development hours reaching 133 would not be disastrous, and whether (this being so) the prospect of consuming only 73 hours makes the prospect a satisfactory one – one better taken up than left alone.

The arithmetic used here is rather arbitrary, and we now face the problem that, although we can examine the balance between advantage and disadvantage in one of the determining factors, there is no way of comparing the influence of different factors.

One way of doing just this is to perform a sensitivity analysis. As its name suggests, this is a way of establishing how big an impact on a project's returns a change of a certain size in one of the input factors can make. For example, in the expert systems project, the decision maker might justifiably want to know what proportion of the project's present value will be lost if the scale of the system turns out to be 20% greater than expected. For it is not unheard of for people to know a lot more, or a lot less, than they think they know.

If an algebraic model can be developed linking driving factors and project returns, the sensitivity can be found mathematically. This can be done in a fairly straightforward way by partial differentiation.[15] But it is most unlikely, in practice, that this will be a realistic approach to take: looking at the driving factors in the expert systems example, it would appear altogether too contrived to build a mathematical model linking net present value and the participants' enthusiasm. This means that there is no real substitute for simple judgement. The analyst would have to gauge as best he could the impact of a shortfall in a factor such as enthusiasm on the costs of the project, and thereby on its net value.

Sensitivities can be defined in a number of slightly different ways. The most straightforward is to say that when a variation of x% from the expected value of an input causes a change of y% in the output, there is a sensitivity of y/x. In some cases this will depend on the value of x, and a sensitivity analysis is in fact normally conducted by finding the impact on the output of a *constant* percentage change in each of the input factors – perhaps 10%. The disadvantage of this is clearly that a 10% variation in different factors may not be comparable, since a 10% variation in one might be pessimistic and in another optimistic. It seems unlikely, for example, that the scale of our expert system could be uncertain by as much as 10%, while we could imagine quite easily that enthusiasm is uncertain by this amount. Where this is the case, an alternative approach is to use two different variations in each input, one being optimistic, the other pessimistic. The rule for selecting the variations would be something along the lines of

> there should be a 5% chance of the variable being greater than the optimistic value and a 5% chance of it being less than the pessimistic value[16]

A pessimistic variation in the expert's ability might be 40%, while a pessimistic variation in the number of users the system will finally have could be just 20%. The problem with this approach is that it more-or-less doubles the workload, and that it re-introduces the need to consider probabilities.

A further difficulty arises because variations in input variables are frequently correlated with one another. In the expert systems case, for example, unexpected variations in the scale of the final system might to some extent be a result of a lack of focus on the part of analysts and users, and a lack of expertise on the part of the project's managers. Saying that a conceivable uncertainty about the manager's expertise can affect the project's value directly by perhaps 30% may, in ignoring its connection with the other factors, be understating its impact. It could affect the output by another 10%, say, indirectly.

Nonetheless, a sensitivity analysis provides a reasonable check of the relative significance of each source of risk, and it demonstrates its absolute significance in terms of its net effects on a project's returns. The information it provides is limited because there is no guarantee that the results of a project will have the same sensitivity to variations in an input over the full range of the input's possible values. But it is a fairly natural approach, and one that is suited to the application of judgement as well as mathematics. A much more complete description of this, and other, ways of analysing risk may be found in a book by Hull.[17]

Mitigating risk

Since risk arises from a lack of information, its mitigation often lies in gathering further information. The risk of suffering an unexpected outcome might be reduced by knowing with greater precision, say, the limits of a system's scale. The analysis described in Section 8.3 suggested a way of calculating how much it would in fact be worth spending on more information, but this assumes that one already has information about probabilities, that probabilities are a good way of making decisions, and that there is no other constraint on finding resources to gather information. Such assumptions do not appear justifiable, and it is quite normal to fall back on intuitive judgements when deciding how far to take the information gathering activity. It is probably even more normal to leave the decision to make itself.

An interesting way of acquiring more information is to develop a prototype. This, in principle, allows a developer to explore the nature of uncertainties at a lesser cost than a full-scale system. In particular, it may help people see more clearly the nature of a system's benefits. It can also avoid some of the revisions that typically have to be made to a system when the specification inadequately captures what the users want.

There are several common-sense guidelines to producing prototypes, mainly developed for application to software systems. An article by Graham,[18] for instance, suggests that prototypes must be both broad and shallow (in order to capture fully the scope of a system), and, in specific areas, narrow and deep (to capture the way in which the system should work in detail). He also recommends that a 'time box' is used to limit the number of iterations the prototype-building goes through. The idea is to force developers and users jointly to set priorities, to avoid the deadly disease of creeping functionality, and to make sure that the prototype is completed in good time.

Prototypes do not, of course, involve the disinterested gathering of facts and figures alone. They are a convenient way of ducking under the threshold values that trigger a full, analytical appraisal. They are also a way of placing a foot in the door: of establishing a system in such a way that it becomes costly to discard. But one can be over-cynical about prototypes. There is often no way of collecting information other than by getting down to a piece of work and reviewing the results, particularly where the work is new and different. Much of the information needed to mould computer systems to the form of the organization they are meant to support is difficult to articulate, and where the computer systems themselves change the form of an organization a prototype can reveal much more accurate information than a purely passive analysis.

There are, in fact, a few ways of mitigating risk that do not involve gathering more information. The first is to transfer the risk, across the market, to another firm (particularly a supplier). When the supplier is better able to deal with the risk because he has more pertinent information, this is the sensible thing to do. Having more information might, in fact, mean that he perceives a lesser risk than his customer. This risk transfer might take the form of fixed-price contracts instead of contracts of indefinite duration at a particular day-rate. Or it might take the form of comprehensive guarantees. For example, a robot supplier could expect to be paid only if the machine he supplies can fully meet the specification he publishes: this specification perhaps speaks about the robot's dynamic performance and its energy consumption. The supplier will doubtless charge for the additional service of assuming risk, but competition will limit the price. And if the supplier perceives a lesser risk in a particular issue the price ought to be less than the cost of the user assuming risk.

When, however, the supplier is less able to manage risk, transferring it to him can rebound on the customer. In most circumstances the customer will suffer as much as anyone if delivery is late, or if there are faults in the design or in the manufacture of the equipment. The supplier might simply fail to meet his contractual obligations entirely, and it is unlikely that the customer will be completely compensated for his lost opportunity costs. Consequential damages, of course, are rarely recoverable.

It is occasionally possible to buy insurance against specific risks from a specialist insurer who balances the risks he accepts in a large and diversified portfolio. But in practice firms are very limited in what they can insure against – the main limitation probably being the presence of moral hazard. This is the difficulty of distinguishing the actions of the insured from uncontrollable events in his environment, and it tends to dull his incentives to take the correct actions.[19] If a firm insures against the possibility that its new CAD system has an NPV of less than £1M then it has relatively little incentive to put any effort into achieving such a return. Since the insurance company would find it hard to say whether a failure to achieve a £1M NPV was due to the firm or to circumstances beyond its control, there would be every temptation for the firm to lie back and claim on the policy. The result is that insurance is not normally a plausible option for a manager attempting to reduce the risk of new technology.

A final, seemingly straightforward, way of mitigating risk is to avoid making a decision entirely. Needless to say, this also tends to rebound on the non-decision maker. In a changing environment there are plenty of uncertainties for firms that attempt to stand perfectly still. And to avoid making a decision is really to make another decision – a decision *not* to evaluate opportunities and *not* to assess threats. It is, in other words, a highly impoverished decision.

Managing risk

The activities of monitoring and acting upon perceptions of risk have no especially arcane characteristics. It is largely a process of understanding the sources of risk, continually re-assessing whether such risks are turning into substantial variations to the project's expected performance, and doing what needs to be done to reverse such variations. Boehm suggests that project managers focus on the top ten or so risks and track them during a succession of review meetings.[20] These meetings examine the progress that has been made toward resolving each of these risks, noting how long they have in fact been a member of the top ten. The ranking is also reviewed at these meetings, and priorities are re-ordered if the circumstances dictate.

To some extent, the organizational processes that are conducted to assess and sanction investments in the first place help to reveal risk, although this may occur in a rather tacit manner. Throwing project proposals backward and forward from one department to another, and opening proposals up to scrutiny, can help locate additional information about the prospects of success. It can reveal bias – whether intentional or not – and it might help to crystallize out any ambiguity about the commitment of potential participants in a new development. This of course suggests that an appraisal shouldn't be kept confidential to a very small number of people – that its reading list should be compiled on the basis of potential interest rather than a strict need to know.

Sometimes the close identification of a project with an individual's career prospects can shake out his true feelings for the viability of marginal proposals, and the bargaining that generally goes on in an organization provides a test of the value a sponsor really attaches to a project. This is not necessarily a good thing, of course, if the sponsor's motivations diverge significantly from those that are formed by the firm as a whole (like increasing present value). It also militates against the sort of risk-taking a commercial firm needs to perform. In particular, there is a danger that what is novel, rapidly changing and less than concrete in its benefits may be ignored. Advanced manufacturing technologies, of course, often display such characteristics.

8.6 Summary

Since discount rates incorporate a project's level of *non-diversifiable* risk, there is a mechanism in the NPV calculation that records an analyst's lack of knowledge about the future effects of an advanced technology. But this is too specialized to say enough about the kinds of risk relevant to technologists and managers. We can therefore think of a process of assessing *total* risk as placing an additional filter on proposals for new developments. Not only do these proposals need to display a positive economic value (a present value and an option value), but they must also satisfy industrial managers that they aren't associated with an excessive risk.

The simplest methods of analysing risk are applied in situations where it is possible to identify

- all possible future states of the world (but not the probabilities that they will occur);
- a set of distinct, alternative courses of action; and
- the pay-off when the future reaches a particular state given that a certain action was taken.

There are then various ways of finding the best course of action according to one's idea of a good decision-making strategy. It is possible, for instance, to find the action that maximizes the minimum possible pay-off.

Although this kind of analysis calls for relatively little information, it is by no means inevitable that this information will be available. With new technology it is often difficult to know in advance of all the properties of the world that bear on a development's value. Therefore one can't necessarily identify all relevant future states of the world.

A more complex analysis is possible when information about probabilities is added, and, in particular, it becomes possible to speak about finding the expected values of each available course of action. Obviously enough, a decision maker would then be able to choose the action which maximizes expected value. But, with probabilities, there is an additional parameter that looks like a good, direct measure of risk – the variance of the probability distribution. Choosing the most desirable action then involves trading off increasing risk (or variance) against increasing returns (or expected values).

Many real situations involve conditions somewhere between the two just mentioned. Perhaps an analyst can say which future states are more likely than others, but he might not be able to say by how much their probabilities differ. The analysis of this kind of model is unfortunately rather complicated, and it involves making a number of assumptions about the decision–making strategy that the model's user will adopt.

There are, in any case, several reasons why the ideas of probability and expected values seem unable to capture what we mean by uncertainty, risk and expectations. We need, as a result, to be wary of using them as a way of making decisions. A plausible alternative is based on the notion of potential surprise: of the surprise people will associate with events unfolding in a particular way. This provides an attractive way of thinking about novel technologies because it draws attention to focus outcomes – that is, to the greatest conceivable losses and gains. These are especially prominent with new systems because the spread of outcomes between them is large. In other words, most people wouldn't be surprised about large deviations from expected outcomes when applying a new technology. Because this spread is large, the centre of one's expectations assumes less importance than it would otherwise.

In practice, it is often sensible to use a number of simple tools and checks to reveal the nature and sources of risk that a firm faces in pursuing a particular development. Sometimes these reveal the information needed to build actuarial or potential surprise models, but they are often a good qualitative filter on their own. An analyst might, for example:

- compile a list of the assumptions underlying his predictions of a project's benefits, and attach a measure of confidence to each of them;
- draw a chart illustrating the primary determinants of each of the main elements of a development's costs, and again attach measures of confidence;
- think about how sensitive the development's returns are to his assumptions about benefits and the levels of the cost drivers.

There are finally, various ways of mitigating risk. A firm can:

- place it on suppliers;
- buy risk-taking facilities from insurers (rather a theoretical option);
- gather more information (for example by building a prototype); and
- avoid risky activities such as introducing advanced technology altogether.

Notes and references

1. Alchian, A. A. Uncertainty, evolution and economic theory. *Journal of Political Economy*, **58**, 211–21 (1950)
2. Burns, T. and Stalker, G. M. *The Management of Innovation*, Tavistock Press, London, p. 5 (1971)
3. Moss, S. J. *An Economic Theory of Business Strategy*, Martin Robertson, Oxford, p. 32 (1981)
4. See Furnham, A. and Lewis, A. *The Economic Mind. The Social Psychology of Economic Behaviour*, Wheatsheaf, Brighton, p. 189 (1986)
5. Pruitt, S. W. and Gitman, L. J. Capital budgeting forecast biases: evidence from the Fortune 500. *Financial Management*, Spring 46–51 (1987)
6. For example, McKenna, C. J. *The Economics of Uncertainty*, Wheatsheaf, Brighton, p. 10 (1986)
7. Petty, J. W., Scott, D. F. and Bird, M. M. The capital expenditure decision-making process of large corporations. *Engineering Economist*, **20**(3), 159–72 (1975)
8. Properties such as stochastic dominance of various orders. See, for example, Barua, S. K. and Srinivasan, G. Investigation of decision criteria for investment in risky assets. *Omega*, **15**(3), 247–53 (1987).
9. Buehler, W. Capital budgeting under qualitative data information. In Crum, R. L. and Derkinderen, F. G. J. (eds.) *Capital Budgeting Under Conditions of Uncertainty*, Martinus Nijhoff, Boston, pp. 81–117 (1981)
10. Carter, C. F. On degrees Shackle: or, the making of business decisions. In Carter, C. F. and Ford, J. L. (eds.) *Uncertainty and Expectations in Economics*, Blackwell, Oxford, pp. 30–42 (1972).
11. This idea is due to Shackle and the remainder of this section draws on his book: Shackle, G. L. S. *Expectations in Economics*, Cambridge University Press (1949)
12. Carter, C. F. *op. cit.*
13. In fact an adjustment ought to be applied to the focus outcomes to reflect the fact that they occur with slightly different levels of surprise. The treatment given here is therefore rather naïve. See Shackle, G. L. S. *op. cit.*, p. 22

14 Egerton, A. Acceptable risk. In Carter, C. F. and Ford, J. L. (eds.) *Uncertainty and Expectations in Economics*, Blackwell, Oxford, pp. 58–73 (1972)
15 Brown, R. A. Sensitivity analysis of capital recovery with return. *Engineering Economist*, **27**(3), 233–7 (1982)
16 Hull, J. C. *The Evaluation of Risk in Business Investment*, Pergamon Press, p. 19 (1980)
17 *Op. cit.*
18 Graham, I. Structured prototyping for requirements specification in expert systems and conventional IT projects. *Computing & Control Engineering Journal*, March, 82–9 (1991)
19 Arrow, K. Economic welfare and the allocation of resources for invention. In Rosenberg, N. (ed.) *The Economics of Technological Change*, Harmondsworth, Penguin, pp. 164–81 (1971)
20 Boehm, B. W. Software risk management: principles and practices. *IEEE Software*, January, 32–41 (1991)

9 Conclusion

A tailor cannot cut the cloth arbitrarily if he wants the suit to fit the customer. Of course, that does not logically exclude the possibility of cutting the cloth arbitrarily, and then slicing up and resewing the customer to fit the suit.

Paul Heyne *The Economic Way of Thinking*

This final chapter draws the discussion to a close by clearing up a few loose ends. The first section is concerned with the way that the appraisal process ought to be connected with the reporting process, and with the difficulties that this involves. The second section considers some of the issues associated with capital spending budgets. The third takes a quick look at how cost-benefit analysis has been applied to decisions involving social issues – to see if there are any promising elements that can be carried across to advanced technology investments. And the last section reflects on the use of the appraisal yardsticks discussed earlier, examining in particular where their limits lie.

9.1 Measuring results in retrospect

Since we have been concerned with appraisal rather than accounting, all the yardsticks described in earlier chapters are directed at helping people form a quantitative, if uncertain, picture of the future. It is, however, worth considering the opposite side of the coin: the way in which managers and technologists can be given a consistent and informative picture of the past.

Present value recap

Chapter 5 described a process in which the worth of a proposed development was measured in terms of its net present value. To this was added a growth option value, if its calculation appeared to be feasible. Where the result was negative, it was to be set against a statement of unquantifiable benefits for an entirely subjective assessment of whether it was to be worth sacrificing the known, negative economic value in order to experience these benefits. The net present value was found by accumulating the cash streams that would be caused by investing in the development, after they have been discounted to reflect the loss of value accompanying uncertainty and the passage of time. The growth option value was found by calculating the present value of succeeding developments, and adjusting this to reflect the fact that they represented options rather than commitments.

If we knew nothing about the systems that already existed to measure financial performance in retrospect, we would probably expect such systems to reflect the way in which we predict financial performance in advance. A manager would probably feel hard done by if told to act on a forecast prepared under one set of rules, and held to account on a measurement conducted under an entirely different set of rules. He would also find it difficult to adapt his decisions to experience – to use performance measurement as a feedback to his decision making.

Our prescription for a system that measures the results of investing in new developments would be quite a simple one. Suppose the measurements were carried out at time intervals of T (maybe once a year). To understand the performance obtained in the period following a time t we would want to know how much more our firm was worth at time $t + T$ than it was at t. Perhaps the increase in worth is a result of cash flows – a new development making appreciable savings or increasing turnover. It might, on the other hand, come from providing more capital value (that is, earning power for the future). Many developments will provide a combination of immediate earnings and future, potential earnings. Therefore our performance measure should incorporate both the cash flows and the increase in future earning power (EP) observed in a particular period. It should be something along the lines of:[1]

(net cash flows between t and $t + T$ plus closing EP of assets) minus

(opening EP of assets uplifted by a discount rate)

The earning power, EP, is the net present value plus the growth option value of our assets. In other words, it represents the value of the future cash streams that the assets will, or will optionally, generate. The opening and closing EPs denote earning power at the beginning and the end of the measurement period, respectively. The net cash flows represent the earnings made during the period and, in the absence of any replenishment, their generation would normally reduce some of the assets' remaining earning power. To uplift the opening EP is simply to reflect that this is measured T units earlier than the closing EP: to make them comparable one of them has to be adjusted by the discount rate. Since T is relatively small, we hope, the discount rate shouldn't be critical. We might use a company-wide value rather than make the specific project adjustments described in earlier chapters. (It is something of an approximation not to uplift the cash flows, of course, unless they happen to occur at the end of the period. But this will matter little if, again, T is quite small, and if the period's cash flows are substantially less than earning power.)

Suppose that in a specific instance we make T equal to one year, and the discount rate 20%. We might have the following schedule of historical measurements (in £Millions, say):

Year	1	2	3	4	5	
(*a*) *Opening EP*	1.0	1.5	2.0	3.0	3.5	
(*b*) *Closing EP*	1.5	2.0	3.0	3.5	4.5	
(*c*) *Uplifted opening EP*	1.2	1.8	2.4	3.6	4.2	[= (a) × 1.2]
(*d*) *Net cash flows*	0.2	0.3	0.3	0.0	0.1	
Performance	0.5	0.5	0.9	(0.1)	0.4	[= (d) + (b) − (c)]

Here, we evidently have a steadily rising stock of nominal earning power, which perhaps reflects the action of inflation and a small growth in the firm's capital base. Cash flows are favourable except in Year 4. The impact of discounting is also Apparent in Year 4: if the need to discount had been ignored, the firm would have appeared to have performed reasonably well, not badly, during that period.

This approach lets us assess the performance in each period, as it is completed, in a way that cannot ignore long-term prosperity. If we simply measured earnings (as net cash flows), and ignored earning power, we could apparently boost performance by liquidating assets. Here, however, the loss this brings for the future would be clear to see as a reduction in the closing EP.

Also, since assets are expressed in terms of their earning power, or NPV plus GOV, it ought to be easy to demonstrate how new developments affect our performance measure. If, for instance, our firm came upon a scheme in Year 3 that promised an NPV of £0.2 million, then this would simply add £0.2 million to the closing EP of Year 3. It would increase our net performance for that year by the same amount because we have, effectively, added this much to the earning power of the firm. As the new development became established in the firm's operations, it would gradually lose its present value, and, at the same time, draw in cash flows. The extent to which it did so could be compared with the schedules drawn up at the time the development had been proposed.

So not only do we have a way of measuring the absolute worth of something, we have a way of testing our forecasting abilities – of calibrating the judgement we exercise when predicting in advance cash flows and earning powers (which are themselves just accumulated future cash flows).

Real measuring systems

In practice, measuring and reporting systems don't generally work in this fashion. Financial accounts (which are directed towards people outside a company) place its performance against yardsticks that differ substantially from those we have applied to investment appraisal. They are based, for example, on the idea of *matching* financial flows: assigning expenses incurred in the production of a product, say, to the period in which the product is sold. Net present value (of course) assigns such changes to the periods in which cash flows take place. Even management accounts, although they are intended to inform decision makers within a company, still don't fit properly with investment appraisal models. In many cases they continue to rely on conventions that no longer reflect a manufacturing world that has begun to move on.

ere the options in fact taken up? Options do not have to be taken up, of
urse, and it would be wrong to do so if by the maturity date the optional
elopment had ceased to be valuable. But if options are taken up then they
e obviously realizable.
e the options, if they were taken up, treated as options rather than
mitments? Was the decision to adopt them, in other words, quite separate
the initial decision that gave rise to the options?
the options such that the initial development was an essential prelimi-
or did it simply enhance the worth of subsequent work? (In other words,
he option on an entire project, or on a lesser parcel of extra value?)
the options exclusive to the firm, or were they pretty much open to all
in the same industry? The value of the options would probably have
reater if the project had been an essential preliminary to a set of exclusive
pments. Answers to such questions will at least suggest whether initial
ts were in the right region.

iled discussion of project reviews and accounting systems would be
here, but there are clearly difficulties in bringing the processes of
nd measurement into line with one another. These difficulties must
faced because appraisal is not an isolated activity, and the tests of
isal system are not limited to self-consistency. It somehow has to slot
administrative structure.

and budgets

t present value discussed in Chapter 5 are based on the separation
isions and financing decisions. There is no need to know how a
pment is to be funded to say whether or not it is worthwhile.
assumed (again implicitly) that firms with potential developments
net worth would always be able to find funding to exploit them.
ognition, in other words, of predetermined limits to the scale of
uld be called on.
is not of course the case. Raising money is costly and it tends
irregularly. More normally, the resources available will be
irm's earnings and the proportion of them it feels it can retain.
be an entirely arbitrary limit. There are analytical ways of
ital rationing in the appraisal process – using mathematical
niques, for instance. They can equally be applied in cases where
an money are rationed, such as the supply of technological
approaches make assumptions that are often difficult to justify.
e, for instance, that the returns a project will make and the
nsume are known beforehand with perfect certainty. The
mainder of this section will therefore be confined to issues of
ure.

We know, for example, that certain ways of apportioning costs can be highly misleading, and that we ought to start paying attention to new-ish ideas about activity-based costing. These should help us to understand much better what determines a firm's costs and whether it is worthwhile continuing to meet them.[2] We also know that the old ideas about capital value have their limitations. High levels of current assets like work-in-process stocks, for example, lead to large holding costs. They are associated with long lead-times and they obscure poor quality production. To speak of them as assets at all is sometimes open to question in anything other than an arid, technical sense.

Fixed assets, on the other hand, tend to be valued according to their purchase costs, reduced by the application of a fixed depreciation rule. It would only be a matter of the purest coincidence if these values worked out to be the same as the earnings stream that this alleged earning power could in fact generate. It is sometimes even convenient to write down the value of productive machines to nothing, for this can help boost the return on capital figures in future periods. It would be quite wrong for a manager to believe his own propaganda, however, and to act as though machines with earning power really had no value.

New technologies can make measurement harder still, and it is a common observation that, when we are trying to determine the effects of something like a flexible manufacturing system, most accounting systems fall well short of the mark. An FMS's effects will most likely range across a number of departments, a number of product lines, the responsibilities of a number of managers, and so forth. Disentangling its costs (those that are *caused* by the presence of the FMS) from everybody else's will be a daunting task: finding its benefits from the output of a traditional reporting system will probably be impossible.

Some practical approaches

There are of course a number of reasons why, in practice, measuring systems have to fall back on procedures and assumptions that aren't justified in principle. Some of these reasons are quite compelling: we ought to be careful of asking for information whose cost exceeds any possible benefit. It might even be the case that old practices and rules of thumb could, in some instances, represent a balance of cost and benefit that is near enough optimal for all practical purposes.

It is therefore worthwhile looking for ways of using accounting results in a way that roughly corresponds to the principles on which investment appraisal is based. The use of an index known as residual income (RI) is one approach that might be suitable. Residual income is found by subtracting from a unit's operating income a notional charge for finance – an opportunity cost of capital – and taking what remains to be a measure of economic performance over the period. It is in fact consistent with the use of net present value provided that the finance charge is levied on capital assets that are valued by their NPV. [3] However, it relies on the use of accounting profits rather than cash flows, which still leaves us with the problem of artificial timings: of costs matched with revenues, and of arbitrary depreciation rules. The use of residual income is similar to the idea of estimating economic rents from accounting figures.[4]

A second approach is known as earned economic income, or EEI.[5] As with other accounting systems incorporating the matching principle, the asset values shown on position statements compiled according to EEI do not express real economic values. (In other words, an asset is not valued according to the cash stream it will generate.) Instead, value is recorded in terms of the past outlays on an asset that haven't yet been set against revenues: that is, in terms of what remains of purchase prices after depreciation. Nonetheless, EEI attempts to make these depreciation charges more realistic indicators of the earning power that is lost during the period of interest. This is done by working out the cash savings that flow from *owning* an asset – as opposed, for example, to renting or leasing it, or buying second-hand assets of the same age. By equating depreciation with this saving, rather than with a direct measure of lost earning power, one avoids having to link the depreciation charge with a particular set of revenues.

Using Grinyer and Elbadri's notation,[5] if the ownership savings in year n of an asset's life are C_n then the fraction of an asset's value written off in year p would be:

$$A_p = C_p(1 + i)^{-p} \Big/ \sum_{j=1}^{N} C_j(1 + i)^{-j}$$

where i is a discount rate. This is the present value of the ownership savings in year p (looking from the beginning of the asset's life), divided by all the savings over the asset's lifetime of N years, also discounted. The depreciation in year p is then simply

$$D_p = A_p I(1 + i)^p$$

where I is the initial outlay on the investment. D_p is the fraction of the asset's value written off in year p multiplied by the value of the asset, uplifted to year p.

Both residual income and earned economic income are intended to be measures of the performance of complete divisions of a firm – in particular, of those whose managers have some discretion over capital investment. (That is, of investment centres.) An aggregated measure of this type is therefore not much help to someone intent on reviewing the outcome of a specific development. To do that, he would have to separate out the effects of that one development from the finances of other activities within the division. This is especially hard when an effect is one of degree rather than one of a distinct kind. And it is much more than simply a question of comparing financial measures before and after the implementation of the development. It is also a question of distinguishing the effects of one development from another, and of distinguishing them from outside influences (such as the prevailing economic climate) – from the background noise if you like.

This probably means that, for the most part, single developments have to be reviewed by referring to a structure of measurements constructed specifically for that development. This structure *may* use values for certain quantities obtained from a firm's accounting systems, but it will combine these values in a unique way. In fact, it is often sensible to use the structure of the initial project appraisal to carry out the retrospective review, substituting actual cash flows for expected

cash flows. Even where it is impossible to find the actu
management accounts, it is still sometimes reasona
simply by asking whether the words used to justify i

In other cases, one can test operational characterist
appraisal of an expert system, for instance, might ha
cash flow £100000 – on the basis that it saved an ex
hours addressing routine problems. Unless there
opportunity cost of an hour of this expert's time h
appraisal and date of review, it is enough to ask i
time has been saved. It isn't necessary to look a
salaries bill that have been apportioned to the ar

We have, as yet, given no recognition to grow
supported on the basis of predicted growth o
attempt to measure the actual growth option
gone ahead? The answer would certainly be y
and inspired a reasonable degree of confidence.
in Chapter 6, growth options are far from
parameters that determine their value are trick
of principle when we apply the option pricin

For example, one of the option paramet
the value of the underlying asset over the p
We have, assuming that the value move
somehow to assess the variance of this ran
is just as acute in retrospect as it is in ad
shares it is easy, of course: you can simpl
period and work out the variance from
question is the opportunity to invest in a
management system this is plainly not p
6, it is often difficult even to identify a

Since this is a practical book, the us
any further. But regardless of the fa
because of practical limitations, the o
way of calibrating, if you like, the
process. It is therefore important to
the quality of our growth options a
done is to apply a number of info
development:

(1) Does it appear, in retrospec
other words, were they genu
and did the firm have the r
out to be worth exercisin
integration of applications
of a local area network,
would have prevented the
had been about to disca
likely components of an

The rationing of resources that is applied over a particular period is usually expressed in the form of a budget. Budgets limit the extent to which people need to search for new developments since they restrict the number of allowable options according to the resources they consume. Another way of looking at this is to say that a given proposal has now to satisfy at least *two* tests if it is to have a chance of being adopted: it must both promise a positive economic value (net present value plus growth option value), and promise a consumption of resources that falls within a certain limit.

It is normal in fact that budgets recognize events only over a single, forthcoming year. It is also typically the case that a distinction is drawn between capital and revenue spending. Capital budgets are applied to resources that will be consumed to produce revenue in future periods, while revenue budgets cover resources that are notionally consumed to produce revenue entirely within the budget period. Capital typically includes machinery and, sometimes, software: revenue items are normally recurring expenses. To make life easier for administrators, the division between the two is more often based on a monetary threshold than on a pattern of use. Revenue budgets might therefore, in practice, apply to all transactions of less than, say, a thousand pounds.

The way in which budgets are managed commonly raises one or two problems. The first is that capital budgets which look one year ahead discourage the forward planning needed to sustain work that extends over several years. Control can only be exercised by referring the amount of money spent to a part of a project, not the whole. This makes it perfectly possible to increase costs over the complete project in the process of reducing costs in a specific period. Budgets also tend to even out the natural lumpiness in capital spending – the fluctuations from one year to the next that stem from the fact that the purchase of things such as computers and computer programs can't be divided down into small parts. This interferes with the ability to follow strategies.[6] It is hard to pursue a pattern of interconnected developments when they have, individually, to be postponed or brought forward in order that their costs in aggregate fit within certain boundaries. Budgets can, you might say, contradict natural patterns of precedence.

The division between capital and revenue budgets creates well-known difficulties for people trying to co-ordinate the introduction of small systems such as personal computers. Because such systems are cheap to buy, and because the capital–revenue division is based on monetary value, they are often classified as revenue items. As a result they tend to be assessed less rigorously. They do not to go through a formal appraisal process, and they are not tested against the firm's plans for the development of its systems. Ordinarily, it seems, a firm's managers will lay down quite rigid rules about whose PC you must buy, and whose programs you may run on it; but they will often not bother to establish how the PC will exchange with other systems the information it generates and consumes. Nor will they produce a rationale for the system that is consistent with that of other developments. In other words there is a concentration on technical minutiae instead of a concern with the manner in which the PC fits into the fabric of the firm's systems and operations.

The way of getting around such difficulties is of course to change the manner in which organizational systems discriminate between different types of development. One has to bear in mind that budgets in general, the capital–revenue division and the expenditure threshold are all ways of tackling uncertainty. It wouldn't be sensible to design new procedures based on the idea that obtaining information is a costless process. It is likely that simple, discriminating tests of some sort will have to be retained in order to separate proposals that need substantial attention from those that do not. But it could well be sensible to modify these tests. In place of a system that divided investments between those that exceeded and those that fell short of a thousand pound expenditure, a better system would divide investments that affected the nature of a firm's operations from those that did not. For instance, investing money in a solid modelling application would change the way a drawing office worked in a way that spending money on replacement disk packs would not. One would expect to undertake a more profound analysis of the first than of the second.

Any way of dividing real developments into just two or three categories for the purpose of forming simple organizational rules will inevitably have anomalies of some kind. It might even be the case that such rules are no longer sensible in any form, and that in future the decision that is made about how to make decisions is less automatic than in the past. Perhaps a personal judgement, rather than a simplistic code of practice, ought to determine how rigorous an appraisal should be.

9.3 Social cost-benefit analysis

The costs and benefits of investments made in the public sector are often characterized by their intangibility, their diffuseness and the time that elapses before they are experienced. This has led to suggestions that the disciplines commonly applied to this type of investment can also be used to assess proposals for new manufacturing technology, which shares some (if not all) of these characteristics.[7] The purpose of this section is therefore to explore whether there is enough in common to make it worthwhile attempting to apply the methods of one to the problems of the other.

Cost–benefit analysis is mostly very similar to the approaches described in earlier chapters, particularly the chapter on net present value. It is a way of applying a single, consistent yardstick that will fully capture the notions of cost and benefit recognized by the people affected by an investment. There are, however, some significant elements in a social analysis which wouldn't normally be found in a commercial appraisal, and it is as well to start out by mentioning these.

Social elements

Social cost–benefit analysis is less concerned with the net gain that one distinct entity such as a firm or a person experiences from a new development, and more with the gain to a society. The usual basis for saying that something is worthwhile to a society, as a whole, is that those who benefit from it can compensate those who lose out and still be better off than they were before. This reflects the fact that developments inevitably damage the interests of some while enhancing the interests of others. The worth of a development depends on what is left over when one is set against the other. It is important to understand, nevertheless, that this compensation criterion is only *potential*: it does not stipulate that the gainers do, in fact, compensate the losers, only that they are able to – and continue to profit.

In real developments the gainers will probably not voluntarily compensate the losers, and it may well prove too inconvenient to coerce them into doing so. Imposing a tax, for instance, would be complicated by all the other issues that taxation brings in train.[8] This means that, in practice, developments affect the way wealth is distributed. By investing in the infrastructure of suburbia, for instance, there would be a tendency to transfer wealth to suburbanites from everybody else.

The danger of partiality towards one part of society also arises from the use of market prices as a way of understanding the costs attached to the resources that a new development will consume. The price that people or groups are willing to pay for a commodity, as it is observed in markets, does not simply reflect their *desire* to acquire that commodity; it also reflects their *ability* to acquire it. An impoverished person may want something much more intensely than a wealthy one, but that plainly doesn't imply that he will pay more for it. This means that if decisions are based on observing what people are willing to pay for a commodity, the people with the most wealth will have a bigger vote, as it were. Their spending habits – their preferences – will be reflected in the decision to a greater extent than those of the poor.

A feature of social cost-benefit analysis is that it is concerned with whether the particular redistributions and priorities accompanying a project are a good or a bad thing. If necessary, additional weightings can be introduced in order to reflect specific (and hopefully explicit) social priorities.[9] However, these are issues that are ruled out of the scope of commercial appraisal. Here, we are not concerned with any input costs other than purchase prices (and opportunity costs for services supplied within the firm), nor with any output benefits other than sales revenues. Thus when we are considering the effects of applying advanced technology, we don't attempt to adjust a project's cash flows to reflect the moral or aesthetic quality of their source.

Intangible benefits

There are, however, several aspects of social cost-benefit appraisals that do bear comparison with new technology developments in commercial firms. The first is that it is difficult to forecast cash flows where there is no direct mechanism for putting a price on a relevant commodity or service. This usually means that there is no market for that good: people don't come together to exchange parcels of it for something else (particularly money). Forecasting is not necessarily straightforward even where there is a market, because material changes inevitably occur in the market between the time of appraisal and the time at which the effects of a project are experienced. But forecasting is especially difficult in instances where there is not even a guide to prices at the time of appraisal.

A social good that isn't directly exchanged in a market might be an unspoilt view across open countryside. An advanced technology good might be the prestige and morale that the technology brings to firms that manage to introduce it successfully. Although neither good is traded in its own right, each clearly influences the attractiveness of goods that *are* traded, and this suggests how such things might be valued. An unspoilt view, for instance, is something that will contribute towards the value of houses that can offer it, and we can consider that the extent to which a house's views are unspoilt is one of several attributes that together determine its price in the housing market. Equally, a firm's prestige and morale are doubtless some of the many factors that contribute towards *its* market value.

It is, in principle, possible to separate out a price for each of these attributes by statistical manipulation, and thereby find an implicit value for such things as unspoilt views. Unfortunately, there are probably too many assumptions underlying this process for it to be a very satisfactory one.[10] And there are a number of additional problems when we try to carry the same approach across to the assessment of technology benefits in commercial firms. We would have to find cases where we can get adequate information on existing applications of new technology and on the valuation (typically the stock-market valuation) of the firm once this technology is established. We would have to assume that these cases did not contain relevant elements that could not be reproduced elsewhere. We would also have to take it that the technology was applied sensibly – in a manner that remains appropriate. The idea of using historical information to price the benefits of things so radically new as advanced information systems makes this approach an especially hard one to apply.

Heavy discounting

Another similarity between many public sector developments and advanced manufacturing technology is the length of time it takes for some of the costs and benefits to show themselves. A typical instance in the public sector is that of the decommissioning costs of nuclear power plant. These are incurred some decades after the initial investment in construction and it may be an entirely different generation that has to meet them. An instance in the area of new technology might be the benefits that new systems can have in improving customer retention – perhaps by reducing commercial transaction costs and creating the goodwill that stems from better delivery performance. This benefit will only become evident beyond the cycle over which customers typically switch from one supplier to another. If the rate of customer turnover were only one every three years or so, it would take rather longer than this to be confident that the rate had been reduced.

One problem that this introduces is an increased uncertainty about the scale of such delayed costs and benefits. Generally speaking, the farther in the future an effect is expected, the less confidently both its magnitude and timing can be predicted. The yardsticks discussed in earlier chapters have provided a way of recording this uncertainty and of making it a material influence on investment decisions. We also have ways of managing the practical consequences of uncertainty, and we know how to mitigate it to some degree by collecting more information. Beyond that we have simply to accept uncertainty for what it is, and recognize it as an important and interesting element of administrative processes.

Another problem that arises with long delays, however, is that they seem to bring into question the validity of discounting. Discounting attenuates effects far into the future to such an extent that it sometimes seems as though they can be ignored entirely. This happens even when it is apparent, intuitively, that they remain significant. Suppose, for instance, that we were attempting to estimate the present value of the decommissioning of a nuclear power station some twenty years from now. Using a discount rate of 20% means that every £1 million of these costs has a present value of £1M/(1 + 20%)^{20}, or just £26000: using a discount rate of 10%, it would be £149000. It doesn't appear reasonable that we should be so myopic as to reduce some very important issues in the future to such insignificant numbers. Even if we were considering the commercial benefits of customer retention in only five years time, every £1M of benefit would still be reduced to £402000 (at 20%) when set against the immediate costs of investing in the appropriate technology. In other words, if we were to act on the results of applying a present value yardstick, we ought *not* to spend £403000 now in order to make £1M in five years' time.

If there is any doubt about the logic of discounting, one only needs to remember how it corresponds to compounded interest payments. Taking the same risk as we would in acquiring the technology to improve customer retention, we could invest £402000 in financial securities, perhaps, and earn £1M in five years time. It wouldn't make sense to discount the £1M we get from an industrial investment to anything other than £402000 in the present. To do so would suggest a randomness in the way we dealt with the time value of wealth and our uncertainty about the future.

So there appears to be a disparity between intuition and the logic of investment appraisal. The method of discounting is reasonable and natural, but when it is applied to certain schemes the answers it finally produces are much less convincing. However, there are a number of reasons why this disparity is not so great as it first seems, and by seeing why this is so we can demonstrate a couple of ways in which the present value yardstick is sometimes mis-applied.

The first thing that needs to be remembered is that the cash flows forecast for the NPV calculation must be expressed in prices of the day – in the prices expected to prevail at the time that cash flows take place. This means, for instance, that if we are concerned with the effects of customer retention in five years' time, then we should be looking at the influence on future (not present) revenues. If a customer is placing business with us now that earns us £1000000, each year, and if we anticipate a constant inflation rate of around 8%, then we might expect that this customer's business will be earning us (£100000 × (1 + 8%)^5), or £147000, in five years' time. It is this latter figure that should provide the basis for the cash flow schedule.

In effect, the rate at which prices increase offsets a part of the influence of the discount rate. If we thought about a good in terms of its current price, we would inflate this to reflect how far in the future we bought or sold it, only to discount it again to a present value. There is a school of thought, in fact, saying that the risk-free rate of interest used in finding the discount rate incorporates current expectations about price inflation. So if the inflation rate were 8%, and the risk-free interest rate 12%, this would imply a rate of 4% for the pure time value of capital. This is by no means a universally-held view, however, and it is best simply to remember that the NPV calculation works with cash flows expressed in the prices of the day on which they are experienced. This generally makes the effect of discounting much less unsettling than it first appears.

A second thing worth bearing in mind is the way in which one determines the risk premium that is embodied in the discount rate. In Chapter 5 it was said that the discount rate is found as $k = R_f + 0.06\beta$. The value of a firm's β was to be looked up in published tables, and then adjusted to find the project's β. This was done by multiplying it by factors measuring the development's revenue sensitivity and its proportion of fixed costs. It was demonstrated in Chapter 5 that if a new development was intended simply to bring about a saving in the firm's fixed costs then the revenue sensitivity, and therefore β, was zero. There would, in other words, be no premium for non-diversifiable risk. In the case of advanced technologies we know that many of their benefits are associated with marketing better products, and that they *do* therefore display a finite degree of revenue sensitivity. But the fact that we are concerned with process technologies means that in some cases it will be realistic to suppose that the risk premium is zero. In such cases we can simply use the risk-free rate of interest as the discount rate. Bearing in mind that we inflate present prices in order to obtain the scale of effects in the future, the extent to which discounting attenuates future benefits is in fact quite slight – perhaps only three or four per cent a year.

Even for developments that are sensitive to revenues (such as the example of customer retention) the latter part of Chapter 5 suggested that flexible technologies keep the scale of this sensitivity below a level typically associated with industrial plant. Flexibility means that the firm *can*, if it has the will and the intellectual resources, reduce those fluctuations in its turnover that stem from changes and uncertainties in general economic conditions. It is the extent of these fluctuations that affects non-diversifiable risk, and the risk premium in the discount rate. The example in Chapter 5 demonstrated the sort of informal rationale that lies behind the choice of an appropriate adjustment to the development's β. If a firm with a β of exactly one adopts a development with around half the normal revenue sensitivity then the risk premium added to the risk-free interest rate is just 3%. If, as before, we knock something off a risk-free rate of 12% for price inflation (say 8%), we still have an attenuation rate in real value of only 7%.

Now in a social cost-benefit analysis one can go quite a bit further in cutting the discount rate down to size. In the first place, depending on the sort of development to which the analysis is applied, people's behaviour occasionally displays very low or even negative discount rates. This appears to be the case when parents invest in their children's education. If the worth of a development was intended to be measured according to the values of such people, then it would be justifiable to use very low discount rates in the present value calculation.

One might also argue that a component of the discount rates that people apply to their decisions about the future is entirely irrational – that it represents a pure myopia.[11] This contrasts with elements that reflect more rational perceptions such as expectations that wealth will increase over time (and that marginal additions to this wealth have less and less importance). If this myopia is truly irrational then there is a case for ignoring it, and for applying a discount rate from which it has been subtracted. The result, again, is that effects far into the future are less attenuated, and that succeeding generations need not suffer from the long-lasting effects of investments made by current generations.[12]

Even if it were possible to gauge the scale of these two effects, however, they are not applicable to a commercial appraisal. Market prices are enough to tell a firm how much its resources cost, and it is capital markets that price capital resources. It is they that price the effect of risk and the effect of time that are reflected in the discount rate. It would therefore be inconsistent to adjust the discount rate to incorporate strictly personal preferences. To do so would be akin to preparing a household budget using a price for bread that reflected what you thought it ought to cost rather than what it does cost (with the first doubtless being less than the second).

Option value

Finally, it is worth mentioning that many highly-regarded social benefits are in fact options. Again, this is rather similar to investments in new technology, whose optional aspects were discussed at length in Chapter 6.

For example, the preservation of natural beauty is rarely thought to be of value only to those people who will definitely look at it. It is also worth something to those who simply want to have the option to look at it, should they feel like doing so at some point in the future. This echoes the idea that having an option to interconnect MRP, process planning and CAD systems (for instance) is a source of value that should contribute towards the decision about installing a computer network.

These options are more valuable than commitments because those that have them can always back out if it turns out not to be worthwhile taking up the asset on which they are based. Options are about acquiring assets in the future, so the value of the assets at the time of acquisition is uncertain. This means that there is a finite likelihood that they will in fact fall in value: as a result, an option to invest in such assets is more valuable than a commitment to do so. It is nicer to have the option of visiting a beauty spot in a year's time than it is to be committed to it. Something even better might turn up.

In Chapter 6 it was apparent that the value of an option increases both with the time to maturity, and with the breadth of uncertainty prior to maturity. And, as with the cash flows of net present value, the asset value is expressed in the prices of the day: those ruling at the time the option matures. Again, therefore, we have an effect that offsets the attenuation of future value that comes from discounting. Far from being reduced by futurity, option values are magnified.

One has to be careful about this line of argument. It does not say that optional effects are more important than committed ones – only that a particular effect regarded as an option is more valuable than a similar one treated as a commitment. Nevertheless, options seem to be important, and it would be a naïve and pessimistic analysis that ignored them.

9.4 Some closing remarks

In this last section, I want to reflect on the ideas that have been described in the core of thc book – the use of present value and option value yardsticks to assess new manufacturing technologies. This is mainly a matter of seeing where their limits lie, seeing what sort of factors set these limits, and knowing under what circumstances these factors are likely to take effect. I will therefore run through some of the more obvious difficulties that using such yardsticks presents, and gauge how far these restrict their applicability.

Information limits

We have already seen that both present value and option value models need a good deal of information about the future if they are to yield useful answers. They will plainly not work properly if the analyst fails to identify and quantify the full set of costs and benefits, because there is then a danger that alternative courses of action will be ranked in the wrong order. Although they make explicit provision for uncertainty, in the sense that they adjust the value of a development to reflect any lack of confidence in its outcome, they still make considerable demands on the analyst to gather information about this uncertainty. This is more of a problem in the option value case, since with present value we can at least re-phrase questions about uncertainty as questions about revenue sensitivity and fixed costs.

The information-collection problem is accompanied by an information-processing problem. Again, it is the calculation of growth option values that is the harder one to perform. It is perfectly possible, of course, to use computer programs to carry out the routine, intensive elements of these calculations, but this doesn't get round the fact that they are relatively opaque. One can estimate the appropriate inputs, and have a certain measure of confidence in them. But because the calculation is difficult enough to be best treated as a black box it is not easy to be happy that the workings of the black box would always run parallel to a more intuitive judgement. This means that the confidence that might be felt in the output (a growth option value) is likely to be a lot less than that attached to the inputs. And in the case of option value there is no obvious way in which to carry out a rough check of the answer the calculation produces.

Partial solutions

A related issue is the need to attempt a solution to the whole of a problem rather than part of it, even if the solution is not demonstrably the best one possible. It is usually very tempting to solve partial problems if their boundaries are set at the limits of what is currently known and understood. A typical case is that of a firm deciding whether to adopt a new technology, when it would make more sense for it to be deciding about both a new technology and (at the same time) new organizational processes. It is easier, very often, to reduce the problem to a technological one because this is a good deal more clear-cut. But solving it, in what appears to be an optimal way, may well mean that the greater gains associated with better organization have been ignored (so there is at least an opportunity cost). In some instances it might be positively damaging to the firm's worth to proceed with a technology without considering what effects it will have on its organization.

Concealing incomparability

Much as it is convenient to reduce different issues to a common scale of value, it is sometimes an artificial and misleading thing to do. This is a criticism that can be levelled at all methods of financial appraisal because, by definition, they are concerned only with the financial effects of a development. It is also a criticism that is frequently heard when the application of the appraisal is a new technology, supposedly because new technology has a number of more diverse facets than old technology.

Theoretical alternatives

It can also be misleading to set likely courses of action against alternatives that are known not to be favoured. The fact that some are more likely than others colours people's views of them, and they apply different criteria when they judge them. Attaching values to choices that are never exercised is, in particular, an arbitrary process: there is a well-known gap between word and deed, and when the word is not followed by the deed there is less reason for it to be a careful and balanced one. This is partly a problem of forming a forgone conclusion, where alternatives are explored simply for the sake of making the appraisal process seem dispassionate. It is also a problem of intuitive judgement simply beating the more methodical system to the answer. Human beings can rarely wait for a full and balanced assessment before reaching an opinion on a particular issue.

Are these problems enough to make the yardsticks invalid or useless? The answer is probably not, for a number of reasons. First, there is nothing *characteristically* difficult about these particular yardsticks in comparison with any others. We are likely to have problems with the payback period or return on investment that are the equal of those with present value. And of course growth option value has no obvious counterpart in the collection of rules of thumb, so it cannot be said to be worse than any of them.

Our difficulties are associated with understanding the pattern of causes and effects that lie between the introduction of a technology and the increase or decrease of a firm's cash flows – not with the way in which the cash flows are manipulated to yield a measure of value. The information problems are therefore probably common to all quantitative methods.

Second, provided that we recognize the incompleteness of what the yardsticks convey, they do not stop us acting on information that they cannot deal with – with what are loosely called intangible effects. We have already seen that it is sensible to find the economic value of quantifiable issues, and then put it beside a statement of unquantified effects to ask whether any shortfall in the quantified value seems to be more then compensated for. We can also use the yardsticks in a weakened role, such that positive economic value remains a necessary, but no longer a sufficient, condition.

This allows us to avoid the problem of optimizing the performance of a sub-system at the expense of the entire system. It doesn't, however, guarantee that we manage to do so. If the explicit, analytical part of an appraisal is incomplete because it fails to capture fully the effects of a technology or the motivations of people, it is equally possible that the more instinctive types of judgement lack consistency and rationality. And sometimes decision makers haven't even a basic understanding of the nature and purpose of a technology they are making decisions about.

Third, the ideas that underlie present value and option value can be used for qualitative reasoning. Although the greatest stress is usually laid on the numerical application of our yardsticks, this is not the only way they can be used. It would be wrong to dismiss them as being invalid because in practice it proves hard to gather all the numbers that they seem to need. We know from the ideas underlying NPV, for instance, that if we were faced with two alternative courses of action with about the same level of non-diversifiable risk then we should compare the patterns of their cash flows. Although timings are important, we can sometimes reason that one option dominates another without quantifying cash flows on a cardinal scale. In other words it will sometimes be enough to know whether a cash flow predicted for one development exceeds that of another in order to know which alternative has the greatest NPV. Since we will always be comparing alternatives (one of which is usually to do nothing), this process of comparison, rather than absolute measurement, is sometimes enough – provided that we consider a reasonable number of alternatives.

To say that two developments have the same non-diversifiable risk is not as hard as it seems, since we need only consider their relative revenue sensitivities and proportional fixed costs. The revenues of many developments will be no more or less sensitive than those of the company as a whole, and, equally, many will make no difference to the proportion of fixed costs a firm has to carry. Moreover, the two factors will generally act in opposite directions (as seen in Chapter 5), and this will tend to reduce the errors we make when estimating the scale of their effects.

Qualitative reasoning is less satisfactory when one alternative is better than another in one respect and worse in another, but even then it will sometimes be reasonable to set one against the other in words rather than figures. We have already seen that there will be effects that just aren't sensibly quantifiable, so we might only be taking a step that we would have faced anyway. This is almost certainly the case with growth options because there quickly comes a point where the inputs to the option pricing model become too blurred to be worth quantifying. But even if we cannot estimate, say, the uncertainty attached to the worth of the optional integration investment following the installation of a computer network, we can still point out that it exists. We can also reason that this uncertainty does not detract from the value of the option. By sticking to informal reasoning, we do not need to know the precise relationship between growth option value and the parameters that determine its value – only the direction in which the relationship acts. We can therefore avoid having to apply formulae whose logic isn't very clear at an intuitive level.

The problem of reducing incomparable factors to a single scale is again one that is common to all quantitative methods. It does not, however, arise from the methods themselves, but from what we are attempting to do with them. The various factors are reduced to a common scale simply because we want to know whether it is worth sacrificing resources to obtain a number of predicted benefits: we have a decision to make, and the things we gain from the decision may be different in kind from the things we lose. Without a basis on which to compare them, we can only make the decision on random or arbitrary grounds. What we must not forget, of course, is that the reduction of these different types of effect to a common scale is only done for the purpose of taking a decision – it has not changed the nature of these effects, and it has certainly not made them any more homogeneous.

It is probably the problem of the foregone conclusion that is really the most difficult to solve. It is very difficult to conduct an appraisal in a manner that does not, to some extent, influence the result on the basis of anything other than the explicit information that goes into it. The fact that decisions are often predetermined is partly evident in the very limited number of alternative developments that are usually appraised (sometimes just one), and in the remarkably small number of developments that reach the appraisal stage but turn out not to have a positive net worth. In other words, the desires and expectations that people form while projects are ill-defined introduce bias to both the search process and the subsequent evaluation process.

It is sometimes said that the ideal is to be a realist when making a decision but an optimist when implementing it: to be dispassionate before a commitment is given, but wilful and determined afterwards. We might wonder whether such an ideal will ever be approached, because it suggests that people should go out and discover the facts, and then become deeply motivated by them. Real people, of course, are motivated by ideas and techniques (especially new ones), and perhaps go in search of facts only as an afterthought. And that is probably the thing that most limits the usefulness of the appraisal process when it is applied to advanced technology.

Notes and references

1. An approach along these lines is described in Allen, D. Strategic financial management. In Bromwich, M. and Hopwood, A. G. (eds.) *Research and Current Issues in Management Accounting*, Pitman, London (1986)
2. One of the more detailed and practical descriptions of ABC in use may be found in Haedicke, J. and Feil, D. Hughes Aircraft sets the standard for ABC. *Management Accounting (US)*, February, 29–33 (1991)
3. Emmanuel, C. R. and Otley, D. T. The usefulness of residual income. *Journal of Business Finance and Accounting*, **3**(4), 43–50 (1976)
4. *The Economist.* Economics focus. Mirror, mirror on the wall. 28th July p. 69 (1990)
5. Grinyer, J. R. and Elbadri, A. E. A case study on interest adjusted accounting using EEI. *Accounting and Business Research*, **19**(76). 327–41 (1989)
6. Currie, W. Managing technology: a crisis in management accounting? *Management Accounting*, February, 24–7 (1991)
7. For example Boddy, D. and Buchanan, D. A. *Managing New Technology*, Blackwell, Oxford, p. 54 (1986)
8. Pearce, D. W. and Nash, C. A. *The Social Appraisal of Projects*, Macmillan, London, p. 30 (1981)
9. Adjustments to reflect national efficiencies and social objectives are described by, for example, Irvin, G. *Modern Cost–Benefit Methods*, Macmillan (1978)
10. For instance, it has to be assumed that the market is a perfect one, and that the different attributes take effect independently of one another: see Pearce, D. W. and Nash, C. A. *op cit.* p. 138

11 Pearce, D. W. and Nash, C. A. *op cit*. p. 154

12 A further suggestion is that, rather than adjust the discount rate downwards, we should increasingly compress time intervals as they extend into the future. See Morley English, J. A perceptual-time scale for determination of a discount function. In van Dam, C. (ed.) *Trends in Financial Decision Making*, Martinus Nijhoff, Leiden, pp. 229–47 (1978)

Index